28. Han, H.; Li, J.; Santos, H.A. Recent advances in Fenton and Fenton-like reaction mediated nanoparticle in cancer therapy. *Biomed. Technol.* **2023**, *3*, 40–51. [CrossRef]
29. Liu, S.; Long, Z.; Liu, H.; Wang, Y.; Zhang, J.; Zhang, G.; Liang, J. Recent advances in ultrasound-Fenton/Fenton-like technology for degradation of aqueous organic pollutants. *Chemosphere* **2024**, *352*, 141286. [CrossRef] [PubMed]
30. Babu, S.G.; Karthik, P.; John, M.C.; Lakhera, S.K.; Ashokkumar, M.; Khim, J.; Neppolian, B. Synergistic effect of sono-photocatalytic process for the degradation of organic pollutants using CuO-TiO_2/rGO. *Ultrason. Sonochem.* **2019**, *50*, 218–223. [CrossRef] [PubMed]
31. Khan, F.; Zahid, M.; Bhatti, H.N.; Jamil, Y. Degradation of persistent organic pollutant using Ag-doped ZnO-ZnS–polyaniline composite as photocatalyst. *Int. J. Environ. Sci. Technol.* **2022**, *20*, 4811–4826. [CrossRef]
32. Rahim Pouran, S.; Abdul Aziz, A.R.; Wan Daud, W.M.A. Review on the main advances in photo-Fenton oxidation system for recalcitrant wastewaters. *J. Ind. Eng. Chem.* **2015**, *21*, 53–69. [CrossRef]
33. Machado, F.; Teixeira, A.C.S.C.; Ruotolo, L.A.M. Critical review of Fenton and photo-Fenton wastewater treatment processes over the last two decades. *Int. J. Environ. Sci. Technol.* **2023**, *20*, 13995–14032. [CrossRef]
34. Merényi, G.; Lind, J.; Naumov, S.; Sonntag, C.V. Reaction of Ozone with Hydrogen Peroxide (Peroxone Process): A Revision of current mechanistic concepts based on thermokinetic and quantum-chemical considerations. *Environ. Sci. Technol.* **2010**, *44*, 3505–3507. [CrossRef]
35. Rekhate, C.V.; Srivastava, J.K. Recent advances in ozone-based advanced oxidation processes for treatment of wastewater- A review. *Chem. Eng. J. Adv.* **2020**, *3*, 100031. [CrossRef]
36. Cao, Y.; Li, J.; Wang, Z.; Guan, C.; Jiang, J. The synergistic effect of oxidant-peroxide coupling systems for water and wastewater treatments. *Water Res.* **2024**, *249*, 120992. [CrossRef]
37. Saharan, V.K.; Pinjari, D.V.; Gogate, P.R.; Pandit, A.B. Advanced oxidation technologies for wastewater treatment. In *Industrial Wastewater Treatment, Recycling and Reuse*; Ranade, V., Bhandari, V., Eds.; Butterworth-Heinemann: Oxford, UK, 2014; pp. 141–191.
38. Elkacmi, R.; Bennajah, M. Advanced oxidation technologies for the treatment and detoxification of olive mill wastewater: A general review. *J. Water Reuse Desal.* **2019**, *9*, 463–505. [CrossRef]
39. Babu Ponnusami, A.; Sinha, S.; Ashokan, H.; Paul, M.V.; Hariharan, S.P.; Arun, J.; Gopinath, K.P.; Hoang Le, Q.; Pugazhendhi, A. Advanced oxidation process (AOP) combined biological process for wastewater treatment: A review on advancements, feasibility and practicability of combined techniques. *Environ. Res.* **2023**, *237*, 116944. [CrossRef] [PubMed]
40. Yu, T.; Chen, H.; Hu, T.; Feng, J.; Xing, W.; Tang, L.; Tang, W. Recent advances in the applications of encapsulated transition-metal nanoparticles in advanced oxidation processes for degradation of organic pollutants: A critical review. *Appl. Catal. B Environ.* **2024**, *342*, 123401. [CrossRef]
41. *ISO 9001:2015*; Quality Management Systems. Available online: https://www.iso.org/standard/62085.html (accessed on 11 June 2024).
42. Priyadarshini, M.; Das, I.; Ghangrekar, M.M.; Blaney, L. Advanced oxidation processes: Performance, advantages, and scale-up of emerging technologies. *J. Environ. Manag.* **2022**, *316*, 115295. [CrossRef] [PubMed]
43. Lee, K.M.; Lai, C.W.; Ngai, K.S.; Juan, J.C. Recent developments of zinc oxide based photocatalyst in water treatment technology: A review. *Water Res.* **2016**, *88*, 428–448. [CrossRef] [PubMed]
44. Antonopoulou, M.; Kosma, C.; Albanis, T.; Konstantinou, I. An overview of homogeneous and heterogeneous photocatalysis applications for the removal of pharmaceutical compounds from real or synthetic hospital wastewaters under lab or pilot scale. *Sci. Total Environ.* **2021**, *765*, 144163. [CrossRef] [PubMed]
45. Fujishima, A.; Honda, K. Electrochemical photolysis of water at a semiconductor electrode. *Nature* **1972**, *238*, 37–38. [CrossRef] [PubMed]
46. Hashimoto, K.; Irie, H.; Fujishima, A. TiO_2 Photocatalysis: A Historical Overview and Future Prospects. *Jpn. J. Appl. Phys.* **2005**, *44*, 8269–8285. [CrossRef]
47. Gaya, U.I.; Abdullah, A.H. Heterogeneous photocatalytic degradation of organic contaminants over titanium dioxide: A review of fundamentals, progress and problems. *J. Photochem. Photobiol. C Photochem. Rev.* **2008**, *9*, 1–12. [CrossRef]
48. Pan, Z.-Z.; Li, Y.; Zhao, Y.; Zhang, C.; Chen, H. Bulk phase charge transfer in focus—And in sequential along with surface steps. *Catal. Today* **2021**, *364*, 2–6. [CrossRef]
49. Bognár, S.; Maksimović, I.; Putnik, P.; Orčić, D.; Putnik-Delić, M.; Šojić Merkulov, D. Integrated approach for sustainable degradation of tolperisone hydrochloride from water by photodegradation: Chemometrics, chemical kinetics, intermediates, and environmental toxicity assessment. *J. Photochem. Photobiol. A Chem.* **2024**, *453*, 115628. [CrossRef]
50. Finčur, N.; Šojić Merkulov, D.; Putnik, P.; Despotović, V.; Banić, N.; Lazarević, M.; Četojević-Simin, D.; Agbaba, J.; Abramović, B. Environmental photocatalytic degradation of antidepressants with solar radiation: Kinetics, mineralization, and toxicity. *Nanomaterials* **2021**, *11*, 632. [CrossRef] [PubMed]
51. Despotović, V.; Abramović, B.; Šojić, D.; Kler, S.; Dalmacija, M.; Bjelica, L.; Orčić, D. Photocatalytic degradation of herbicide quinmerac in various types of natural water. *Wat. Air Soil Pollut.* **2012**, *223*, 3009–3020. [CrossRef]
52. Abramović, B.F.; Despotović, V.N.; Šojić, D.V.; Orčić, D.Z.; Csanadi, J.J.; Četojević-Simin, D.D. Photocatalytic degradation of the herbicide clomazone in natural water using TiO_2: Kinetics, mechanism, and toxicity of degradation products. *Chemosphere* **2013**, *93*, 166–171. [CrossRef] [PubMed]

53. Šojić, D.V.; Orčić, D.; Četojević-Simin, D.D.; Banić, N.D.; Abramović, B.F. Efficient removal of sulcotrione and its formulated compound Tangenta® in aqueous TiO_2 suspension: Stability, photoproducts assessment and toxicity. *Chemosphere* **2015**, *138*, 988–994. [CrossRef] [PubMed]
54. Šojić Merkulov, D.; Lazarević, M.; Djordjević, A.; Nafradi, M.; Alapi, T.; Putnik, P.; Rakočević, Z.; Novaković, M.; Miljević, B.; Bognar, S.; et al. Potential of TiO_2 with various Au nanoparticles for catalyzing mesotrione removal from wastewaters under sunlight. *Nanomaterials* **2020**, *10*, 1591. [CrossRef] [PubMed]
55. Lazarević, M.; Putnik, P.; Šojić Merkulov, D. Chemometric evaluation of different parameters for removal of tembotrione (agricultural herbicide) from water by adsorption and photocatalytic degradation using sustainable nanotechnology. *Food Energy Secur.* **2022**, *11*, e368. [CrossRef]
56. Xu, D.; Cheng, B.; Cao, S.; Yu, J. Enhanced photocatalytic activity and stability of Z-scheme Ag_2CrO_4-GO composite photocatalysts for organic pollutant degradation. *Appl. Catal. B Environ.* **2015**, *164*, 380–388. [CrossRef]
57. Zhu, B.; Xia, P.; Li, Y.; Ho, W.; Yu, J. Fabrication and photocatalytic activity enhanced mechanism of direct Z-scheme g-C_3N_4/Ag_2WO_4 photocatalyst. *Appl. Surf. Sci.* **2017**, *391*, 175–183. [CrossRef]
58. Oliveira, R.A.; Castro, M.A.M.; Porto, D.L.; Aragão, C.F.S.; Souza, R.P.; Silva, U.C.; Bomio, M.R.D.; Motta, F.V. Immobilization of Bi_2MoO_6/ZnO heterojunctions on glass substrate: Design of drug and dye mixture degradation by solar-driven photocatalysis. *J. Photochem. Photobiol. A* **2024**, *452*, 115619. [CrossRef]
59. Yang, X.; Wang, D. Photocatalysis: From fundamental principles to materials and applications. *ACS Appl. Energy Mater.* **2018**, *1*, 6657–6693. [CrossRef]
60. Wang, H.; Li, X.; Zhao, X.; Li, C.; Song, X.; Zhang, P.; Huo, P.; Li, X. A review on heterogeneous photocatalysis for environmental remediation: From semiconductors to modification strategies. *Chin. J. Catal.* **2022**, *43*, 178–214. [CrossRef]
61. Chen, X.; Shen, S.; Guo, L.; Mao, S.S. Semiconductor-based photocatalytic hydrogen generation. *Chem. Rev.* **2010**, *110*, 6503–6570. [CrossRef] [PubMed]
62. Chong, M.N.; Jin, B.; Chow, C.W.; Saint, C. Recent developments in photocatalytic water treatment technology: A review. *Water Res.* **2010**, *44*, 2997–3027. [CrossRef] [PubMed]
63. Sun, W.; Zhu, J.; Zhang, M.; Meng, X.; Chen, M.; Feng, Y.; Chen, X.; Ding, Y. Recent advances and perspectives in cobalt-based heterogeneous catalysts for photocatalytic water splitting, CO_2 reduction, and N_2 fixation. *Chin. J. Catal.* **2022**, *43*, 2273–2300. [CrossRef]
64. Torres-Palma, R.A.; Serna-Galvis, E.A. Sonolysis. In *Advanced Oxidation Processes for Waste Water Treatment*; Ameta, S., Ameta, R., Eds.; Academic Press: Cambridge, MA, USA, 2018; pp. 177–213.
65. Lim, M.; Son, Y.; Khim, J. Frequency effects on the sonochemical degradation of chlorinated compounds. *Ultrason. Sonochem.* **2011**, *18*, 460–465. [CrossRef] [PubMed]
66. Rooze, J.; Rebrov, E.V.; Schouten, J.C.; Keurentjes, J.T. Dissolved gas and ultrasonic cavitation—A review. *Ultrason. Sonochem.* **2013**, *20*, 1–11. [CrossRef] [PubMed]
67. Matafonova, G.; Batoev, V. Review on low- and high-frequency sonolytic, sonophotolytic and sonophotochemical processes for inactivating pathogenic microorganisms in aqueous media. *Water Res.* **2019**, *166*, 115085. [CrossRef] [PubMed]
68. Patidar, R.; Srivastava, V.C. Evaluation of the sono-assisted photolysis method for the mineralization of toxic pollutants. *Sep. Purif. Technol.* **2021**, *258*, 117903. [CrossRef]
69. Theerthagiri, J.; Lee, S.J.; Karuppasamy, K.; Arulmani, S.; Veeralakshmi, S.; Ashokkumar, M.; Choi, M.Y. Application of advanced materials in sonophotocatalytic processes for the remediation of environmental pollutants. *J. Hazard. Mater.* **2021**, *412*, 125245. [CrossRef]
70. Panda, D.; Manickam, S. Recent advancements in the sonophotocatalysis (SPC) and doped-sonophotocatalysis (DSPC) for the treatment of recalcitrant hazardous organic water pollutants. *Ultrason. Sonochem.* **2017**, *36*, 481–496. [CrossRef] [PubMed]
71. Mapukata, S.; Ntsendwana, B.; Mokhena, T.; Sikhwivhilu, L. Advances on sonophotocatalysis as a water and wastewater treatment technique: Efficiency, challenges and process optimisation. *Front. Chem.* **2023**, *11*, 1252191. [CrossRef] [PubMed]
72. Sun, M.; Yao, Y.; Ding, W.; Anandan, S. N/Ti^{3+} co-doping biphasic TiO_2/Bi_2WO_6 heterojunctions: Hydrothermal fabrication and sonophotocatalytic degradation of organic pollutants. *J. Alloys Compd.* **2020**, *820*, 153172. [CrossRef]
73. Ayare, S.D.; Gogate, P.R. Sonophotocatalytic oxidation based treatment of phthalocyanine pigment containing industrial wastewater intensified using oxidising agents. *Sep. Purif. Technol.* **2020**, *233*, 115979. [CrossRef]
74. Du, B.; Fan, G.; Yang, S.; Chen, Z.; Luo, J.; Yu, W.; Yu, J.; Wei, Q.; Lu, Y. Sonophotocatalytic degradation of 17β-estradiol by Er^{3+}-CdS/MoS_2: The role and transformation of reactive oxygen species. *J. Clean. Prod.* **2022**, *333*, 130203. [CrossRef]
75. Ivetić, T.B.; Dimitrievska, M.R.; Finčur, N.L.; Đačanin, L.R.; Gúth, I.O.; Abramović, B.F.; Lukić-Petrović, S.R. Effect of annealing temperature on structural and optical properties of Mg-doped ZnO nanoparticles and their photocatalytic efficiency in alprazolam degradation. *Ceram. Int.* **2014**, *40*, 1545–1552. [CrossRef]
76. Šojić, D.V.; Orčić, D.Z.; Četojević-Simin, D.D.; Despotović, V.N.; Abramović, B.F. Kinetics and the mechanism of the photocatalytic degradation of mesotrione in aqueous suspension and toxicity of its degradation mixtures. *J. Mol. Catal. A Chem.* **2014**, *392*, 67–75. [CrossRef]
77. Ivetić, T.B.; Finčur, N.L.; Đačanin, L.R.; Abramović, B.F.; Lukić-Petrović, S.R. Ternary and coupled binary zinc tin oxide nanopowders: Synthesis, characterization, and potential application in photocatalytic processes. *Mater. Res. Bull.* **2015**, *62*, 114–121. [CrossRef]

Contents

Fabrication of Carbon and Related Materials/Metal Hybrids and Composites (Volume II)

About the Editors

Walid M. Daoush

Dr. Daoush is currently working as an adjunct full professor at the college of science in Imam Mohammad ibn Saud Islamic University (IMSU) in KSA. He is a professor of materials science at department of production technology of the faculty of Technology and Education at Helwan University. He moved from CMRDI in Cairo to Helwan University in 2010 as an associate professor of materials science. He started his work as a research assistant at CMRDI. In 2004, he awarded his PhD from University of Ain Shams in Cairo. Recently, his efforts have been applied to a wide range of materials and he is a co-author of more than 90 published articles in different fields such as metal-supported CNTs, biomaterials, tungsten heavy alloys, metallic foams and biological effect of nanoparticles. He had Research Assistant Professorships at (KAIST) from the beginning of 2007 to the end of 2008, while serving as a Post-Doc at (NTUST) in 2011. In addition, he has been a visiting scientist in several international institutes, including Japan Fine Ceramic Center (JFCC) and Materials Research Institute of Slovak republic, and recently as a Fulbright visiting scholar in the college of engineering at San Diego State University. Dr. Walid is listed in the membership board of KRF, Who's Who in the world 2009, American Chemical society and Materials Research Society. He received the prize of young scientists in the innovations of Materials Science from the Egyptian academy of Science and Technology in 2010. Dr. Daoush has co-supervised on, so far, 20 theses and published over seventy-five articles, one patent and three textbooks. He collaborated with several national and international organizations for start-up projects in Egypt and different countries in Austrian Institute of Technology (AIT), Materials Research Institute of Slovak Republic, TRC of the Egyptian Armed Forces, STDF in Egypt, the University Carlos III de Madrid and recently to the college of engineering at San Diego State University.

Fawad Inam

Based at the University of East London (UK), Professor Dr Fawad Inam has contributed to achieving more than GBP 2.5 million in research and enterprise income from government and commercial platforms. Followed by over 228k individuals worldwide in publicly accessible general engineering outreach social media channels, he has advised over 55 universities and institutes from all the four nations of the UK. As a British Chartered Engineer (CEng), fellow of the Institute of Mechanical Engineers (IMechE), and principal fellow of the Higher Education Academy (HEA, UK), he has formed/strengthened international partnerships, delivered engineering courses and induced institutional higher education practice in other countries such as China, South Korea, Singapore, Malaysia, India, Pakistan, Saudi Arabia, and the Netherlands, to list a few. He is the recipient of the 'Materials World Award' conferred by the Institute of Materials, Minerals and Mining (IoM3, UK) in 2008. Prof. Inam's research and enterprise expertise is primarily focused around improving the performance and functionalities of engineering products using materials science and engineering, in particular nano-induced smart materials

Mostafa Ghasemi Baboli

Dr. Mostafa is an Associate Professor of Chemical Engineering in Sohar University. Before joining to the Sohar university, he worked in University Teknologi PETRONAS (UTP) and the National University of Malaysia (UKM). He has published more than 85 manuscripts in high impact peer reviewed international journals and some book/book chapters in MDPI and Elsevier. His research mostly focuses on desalination of brackish water and wastewater treatment by microbial fuel cells. He is one of the top 2% scientists of the world published by Sandford University in 2021 and 2022.

Maha M. Khayyat

Dr. Maha M. Khayyat is currently a research professor of physics at (KACST), Energy and Industry Sector. Dr Khayyat received her PhD from Cavendish laboratory, University of Cambridge (Oct. 2001–July 2004), studying the mechanical, optical and electrical properties of crystalline and amorphous semiconductor materials (Si, Ge and GaAs) using indentation techniques at both the micro- and nanoscale. Then, she worked as a collaborative researcher at Cavendish lab, where she was introduced to the topic of photovoltaic cells at Microelecronics Research Center (MRC) and Nanoscience Center of University of Cambridge. She worked within a joint project KACST/IBM, undertaking research at T. J. Watson Research Center, USA. She worked on solar cells based on Si nanowires and then started conducting original research by improving nanowire technology and spalling at liquid nitrogen temperature. After this, she spent her sabbatical year at MIT, Nanoengineering group, funded by the Center of Clean Water of Energy of KFUPM. She aimed to develop surface structures that trap photons in thin films of Si. Then, she worked at Poitiers University, Poitiers, France, studying the properties of GaN nanomembranes for LED applications.

Preface

This Special Issue covers various aspects of materials engineering in the topic of the fabrication of hybrid and composite materials. It mainly focuses on the novel developments of new processing methodologies for the fabrication and modification of carbon and related hybrid and composite materials for magnetic, electronics, prosthetic and optoelectronics applications. These composite and hybrid materials are mainly composed of the molecular structure of fullerenes, graphene, carbon fibers and nanotubes as well as metallic nanoparticles of great potential applications in nanotechnology. The Special Issue collected ten research articles describing diverse research topics of different carbon base materials like single-wall carbon nanotubes, carbon fibers, fullerenes, multi-wall carbon boron nitride and graphene hybrids and nanocomposites. The contributions of this Special Issue can be classified into four groups: the best concerns fabrications of new composite and hybrid materials of the different allotropies forms of carbon, such as carbon fibers, carbon nanotubes and graphene; the second group considers the modeling, computational studies and the dynamics of large molecules which opens up possibilities for a simplified description of many complex systems composed of carbon molecules such as fullerenes and carbon Nano-containers; the third group looks at the synthesis and properties of new nanoparticles and hybrid materials with various applications due to their electrical, mechanical, dye removal and water treatments activities; and the last group is strongly related to a discussion on the fabrication of laminated composites prosthetic foot prototyping and their mechanical properties. Some of the articles generally contain the results of calculations and are compared with experimental observations. The results presented in the articles collected in this Special Issue clearly demonstrate that the carbon-based hybrid and composite materials are clearly very important in the field of new and advanced materials. However, to date, lowering the manufacturing cost and expanding the applications of carbon base hybrid and composites remain challenging. The main goal of this Special Issue is to focus on the points mentioned above for the preparation, characterization, computational studies, simulations and properties of advanced carbon composites and hybrid materials facilitating their practical applications in various fields.

We would like to thank all authors whose contributions are included in this Special Issue for their excellent work, and for their inspiring and interesting articles.

Walid M. Daoush, Fawad Inam, Mostafa Ghasemi Baboli, and Maha M. Khayyat
Editors

crystals

MDPI

Article

Electrical and Structural Properties of CVD-Graphene Oxidized Using $KMnO_4/H_2SO_4$ Solution

Jin-Seok Choi [1], Ki-Sik Im [2], Tae-Kyun Lee [1], Yeo-Jin Choi [1] and Sung-Jin An [1,*]

[1] Department of Advanced Materials Science and Engineering, Kumoh National Institute of Technology, Gumi 39177, Korea; choijs1220@kumoh.ac.kr (J.-S.C.); ltk@kumoh.ac.kr (T.-K.L.); dota23@kumoh.ac.kr (Y.-J.C.)

[2] Advanced Materials Research Center, Kumoh National Institute of Technology, Gumi 39177, Korea; ksim@kumoh.ac.kr

* Correspondence: sungjinan@kumoh.ac.kr

Abstract: We report the electrical properties of graphene grown via chemical vapor deposition (CVD-graphene) and oxidized using a $KMnO_4$/dilute H_2SO_4 mixture. CVD-graphene was successfully oxidized without any pores or peeling off from the substrates. When the H_2SO_4 concentration was increased, the electrical resistance of the oxidized graphene (OG) increased. In particular, OG-20 shows a nonlinear current–voltage curve similar to that of a diode owing to direct tunneling through the interfaces between the nanosized sp^2 and sp^3 regions. The changes in electrical properties occurred because of structural evolution. As the H_2SO_4 concentration increased, the number of oxygen functional groups (epoxide/hydroxyl and carboxyl groups) in the OG increased. In addition, a reduction in the average distance between defects in the OG was determined using Raman spectroscopy. Oxidation using a $KMnO_4$/dilute H_2SO_4 mixture results in CVD-graphene with modified electrical properties for graphene-based applications.

Keywords: graphene; oxidized graphene; oxygen functional groups; potassium permanganate; sulfuric acid

Citation: Choi, J.-S.; Im, K.-S.; Lee, T.-K.; Choi, Y.-J.; An, S.-J. Electrical and Structural Properties of CVD-Graphene Oxidized Using $KMnO_4/H_2SO_4$ Solution. *Crystals* **2022**, *12*, 439. https://doi.org/10.3390/cryst12040439

Academic Editors: Walid M. Daoush, Fawad Inam, Mostafa Ghasemi Baboli and Maha M. Khayyat

Received: 4 March 2022
Accepted: 18 March 2022
Published: 22 March 2022

1. Introduction

The chemical functionalization of a graphene surface can easily modify its electrical, chemical, and optical properties. In particular, oxygen functionalization offers metal–insulator transition by controlling the degree of oxidation, reaction sites for the synthesis of other functional groups, and a route for hole doping [1,2]. Thus far, considerable research has been devoted to controlling the metal–insulator transition by manipulating the degree of oxidation through the partial oxidation and reduction of graphene oxide. Jin et al. reported that oxidized graphene (OG) field-effect transistors (FETs) show p-type characteristics for partial oxidation (for 5 min) and insulating behavior for full oxidation (for 60 min) of graphene in air [2]. Goki et al. reported a carrier transport mechanism in partially oxidized graphene (POG) synthesized by varying the reduction time of graphene oxide [3]. They demonstrated that carrier transport in lightly reduced graphene oxide occurs via variable-range hopping, whereas band-like transport dominates in well-reduced graphene oxide. Carrier transport in graphene oxide is intriguing because of the presence of the substantial electronic disorder arising from variable sp^2 and sp^3 carbon ratios. Graphene oxide represents graphene layers with oxygen functional groups decorated on the basal planes and edges. In graphene oxide, most of the carbon atoms bonded with oxygen are sp^3 hybridized and disrupt the extended sp^2 conjugated network of the original graphene sheet. The substantial sp^3 fraction in graphene oxide can transform the material into a semiconductor and ultimately to a graphene-like semimetal.

Although a large number of studies have been devoted to the synthesis of graphene oxide via chemical exfoliation using graphite flakes, such as Hummer's method, there is limited availability of mass-production processes for graphene oxide for electronic applications because graphene oxide is a micro-sized sheet [4,5]. Graphene oxide sheets cannot form uniform and large-area films. Graphene grown via chemical vapor deposition (CVD-graphene) has the following advantages: (i) it is easy to achieve the scalable production of graphene suitable for the mass production of graphene-based electronic devices; and (ii) it can form monocrystalline and polycrystalline graphene [6–9]. In addition, it can be patterned using conventional lithography techniques and can be transferred onto flexible substrates with negligible structural damage.

To oxidize the CVD-graphene, many studies have been conducted using dry oxidation methods, such as ozone treatment assisted by UV irradiation and oxygen plasma treatment [10–13]. However, these dry oxidation methods can easily form nanoscale pits (or pores) in the graphene [14–16]. In contrast, wet-based chemical oxidation of CVD-graphene has rarely been reported. The wet oxidation of CVD-graphene offers several advantages, such as a low-cost process, a wide range of solvents that can be used, and easy control of the metal-insulator transition through the adjustment of the oxidation time [2,17].

In this study, we demonstrate the variation in electrical properties and structural evolution of oxidized CVD-graphene using the reagents used in Hummers' method, which is a mixed solution of $KMnO_4$ and H_2SO_4. The electrical properties of the oxidized CVD-graphene were evaluated using current–voltage (I-V) plots, as a function of the H_2SO_4 concentration, and by calculating the Schottky barrier height (SBH, Φ_b) and the gate bias effect. The electrical properties were dependent on the structural properties of OG, which were investigated using scanning electron microscopy (SEM), X-ray photoelectron spectroscopy (XPS), and Raman spectroscopy.

2. Materials and Methods

Graphene layers were grown on Cu foils using a thermal CVD method. The detailed growth procedure was as follows: (1) initial step at 5 mTorr and 1000 °C under 10 sccm of Ar gas for 1 h; (2) annealing at 1000 °C under Ar 10 sccm/H_2 50 sccm atmosphere for 20 min; (3) graphene growth step at 1000 °C under Ar 10 sccm/H_2 50 sccm/CH_4 50 sccm for 1 min; and (4) final step at room temperature (25 °C) under an Ar/H_2/CH_4 atmosphere. For the transfer of graphene onto SiO_2/Si wafers, graphene grown on the other surface of Cu foils was first removed by O_2 plasma (SPI supplies, Plasma Prep III) at 100 W for 10 s. Subsequently, the graphene/Cu foil samples were coated with poly methyl methacrylate (PMMA). After the Cu foils were etched in a 0.5 M ammonium persulfate solution, the PMMA/graphene samples were floated on deionized water. Finally, the PMMA/graphene samples were transferred onto SiO_2/Si wafers, and the PMMA was removed by immersion in acetone for 1 h.

To measure the electrical properties, the mesa isolation of the graphene sheet with a dimension of 25 × 50 μm^2 was conducted using O_2 plasma treatment. Subsequently, Pd/Au (20 nm/50 nm) metal layers as ohmic contacts and an Au (50 nm) metal pad as an external contact were deposited using an electron-beam evaporator. The distance between the two electrodes was 13 μm. A surface oxidation region with a length of 7 μm was defined using photolithography.

The surface oxidation of graphene was performed using 0.1 g of $KMnO_4$ in a 10–30% dilute H_2SO_4 mixture. The graphene transferred onto a SiO_2/Si wafer was immersed in the $KMnO_4$/dilute H_2SO_4 mixture for 10 min at 25 °C. Graphene oxidized with different concentrations of H_2SO_4 is denoted as OG-n, where n represents the H_2SO_4 concentration. Finally, the graphene on the SiO_2/Si wafer was immersed in DI water for 1 min and cleaned by N_2 blowing.

Optical microscopy (OM) and field-effect scanning electron microscopy (FE-SEM) were used to characterize the graphene layer transferred onto the SiO_2/Si wafer. High resolution transmission electron microscopy (HRTEM) was utilized to investigate the surface of the samples. The chemical bonds of the oxygen functional groups of the samples were investigated using XPS with a Theta Probe AR-XPS system (Thermo Fisher Scientific,

Fabrication of Carbon and Related Materials/Metal Hybrids and Composites (Volume II)

Editors

Walid M. Daoush
Fawad Inam
Mostafa Ghasemi Baboli
Maha M. Khayyat

MDPI

Basel • Beijing • Wuhan • Barcelona • Belgrade • Novi Sad • Cluj • Manchester

Editors

Walid M. Daoush
Production Technology
Helwan University
Cairo
Egypt

Fawad Inam
Department of Engineering & Construction
University of East London
London
United Kingdom

Mostafa Ghasemi Baboli
Chemical Engineering
Sohar University
Sohar
Oman

Maha M. Khayyat
Energy and Industry
King Abdulaziz City for Science and Technology (KACST)
Riyadh
Saudi Arabia

Editorial Office
MDPI
St. Alban-Anlage 66
4052 Basel, Switzerland

This is a reprint of articles from the Special Issue published online in the open access journal *Crystals* (ISSN 2073-4352) (available at: www.mdpi.com/journal/crystals/special_issues/Carbon_Fabrication_II).

For citation purposes, cite each article independently as indicated on the article page online and as indicated below:

Lastname, A.A.; Lastname, B.B. Article Title. *Journal Name* **Year**, *Volume Number*, Page Range.

ISBN 978-3-0365-8729-5 (Hbk)
ISBN 978-3-0365-8728-8 (PDF)
doi.org/10.3390/books978-3-0365-8728-8

Waltham, MA, USA) with the following parameters: monochromatic Al Kα radiation (hν = 1486.6 eV), X-ray energy of 15 kV, 150 W, and a spot size of 400 μm^2. Micro Raman spectroscopy (System 1000, Renishaw, Wotton-under-Edge, UK) with a 514 nm diode laser was used to analyze the structural variation of the OG structures. OM objective lenses (×100 magnification) were used to acquire the Raman spectra. The spectral resolution of the equipment is 1800 1/mm gratings. The electrical characteristics of the OG structures were investigated based on DC characteristics using a system combined with HP4156C (Keysight, Santa Rosa, CA, USA) and a probe station.

3. Results and Discussion

Figure 1a presents a schematic of the graphene/OG in-plane heterostructure used to evaluate the electrical properties. OG regions were formed using conventional photolithography. The length of the OG region is 7 μm, as shown in Figure 1b. From the OM image, the OG region has a lighter color than graphene, possibly because of the oxygen functional groups on the OG [18]. The Raman spectra shown in Figure 1c indicate the formation of OG regions after oxidation using $KMnO_4/H_2SO_4$ solution. The detailed structural changes in the OG are discussed later using XPS and Raman analyses as a function of the H_2SO_4 concentration.

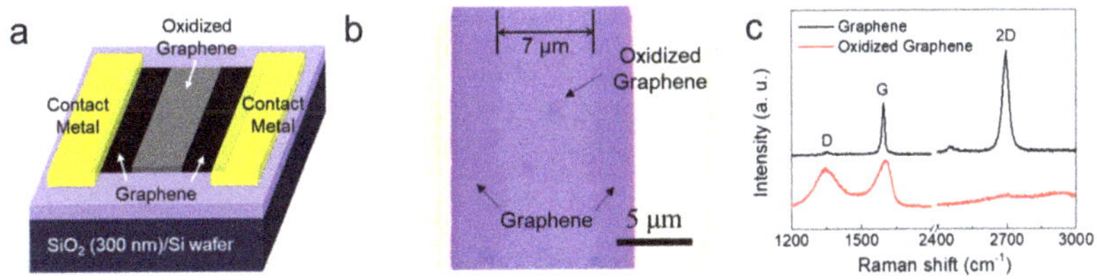

Figure 1. (**a**) Schematic and (**b**) OM image of the graphene/OG heterostructure; and (**c**) Raman spectra of the graphene and OG-30.

Current–voltage (I-V) curves of graphene/OG-10, OG-20, or OG-30 in-plane heterostructures were obtained. As shown in Figure 2a,b, OG-10 displays a resistance approximately 100 times higher than that of graphene. The I-V plot for OG-20 is a nonlinear curve, which is different from that of OG-10 (Figure 2c). The I-V curve of the graphene/OG-20 structure is similar to that of graphene/graphene oxide vertical contact structures [19]. The OG-30 acts as an insulator, as shown in Figure 2d. Figure 3 shows the I_{DS}-V_{DS} curves as a function of the gate voltage. The OG-20 shows p-type behavior. This is due to the presence of oxygen functional groups on the surface and/or edges of graphene [20].

Assuming an ideal diode without shunt and series resistances, the SBH of the graphene/OG-20 in-plane heterostructure can be calculated by employing a thermionic model, using the following Equation (1):

$$\Phi_b = \frac{kT}{q}\ln\left(\frac{AT^2A^*}{I_0}\right) \quad (1)$$

where k is the Boltzmann constant; T is the absolute temperature; q is the electronic charge; A is the contact area of the barrier; A^* is the Richardson constant (119.7 A·cm^{-2}·K^{-2} for a free electron); and I_0 is the saturation current obtained from the I-V curves. The estimated SBH for OG-20 was 0.28 eV.

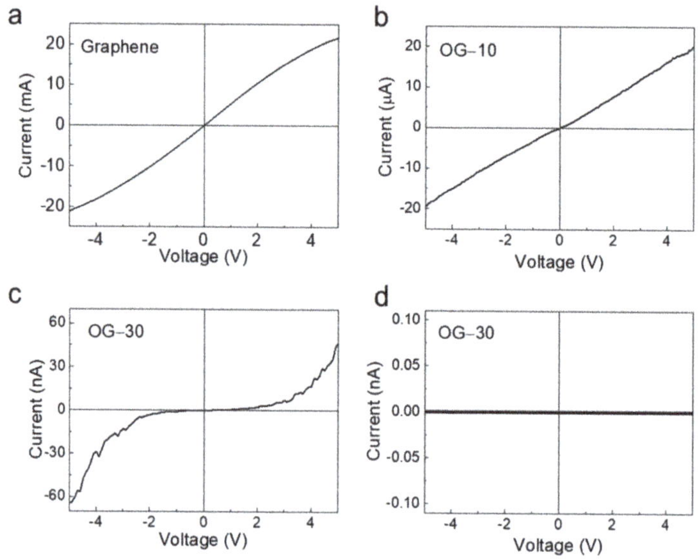

Figure 2. Current-voltage (I-V) curves of (**a**) graphene, (**b**) OG-10, (**c**) OG-20, and (**d**) OG-30.

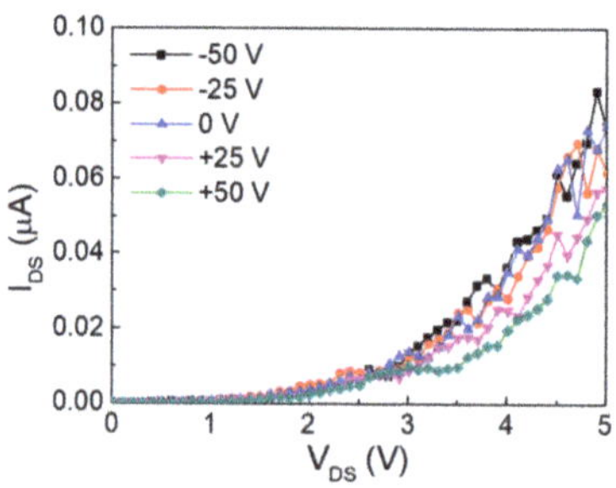

Figure 3. I_{DS}-V_{DS} measurements of the OG-20 as a function of gate-bias voltages.

To demonstrate the change in electrical properties of oxidized CVD-graphene as a function of the H_2SO_4 concentration, we investigated the structural evolution using XPS and Raman analyses. Figure 4 shows the C1s spectra of graphene, OG-10, OG-20, and OG-30. The graphene (Figure 4b) spectrum indicates the presence of strong C–C bonds, a low amount of epoxide (C–O–C)/hydroxyl (C–OH) groups, and a very small amount of carboxyl groups (O–C=O), at 284.6, 285.4, and 288.7 eV, respectively [21]. In the OG-10 spectrum, the amount of epoxide/hydroxyl and carboxyl groups is slightly higher than that in the graphene. Significantly, the peak intensities corresponding to the oxygen functional groups in OG-20 are significantly higher than those in OG-10 (Figure 4d). In addition, when the H_2SO_4 concentration was increased further (OG-30), the number of epoxide/hydroxyl groups increased slightly more than that in OG-20. The peak intensity ratio (I_{OC}/I_{GC}) of oxidized carbon (285.4, 288.7, and 291 eV) to graphitic carbon (284.6 eV) explains the oxidation degree of graphene oxide [10,22]. The I_{OC}/I_{GC} values of graphene, OG-10, OG-20, and OG-30 were calculated as 0.24, 0.3, 0.6, and 0.84, respectively (see Table 1). Therefore, the increase in the amount of oxygen functional groups changes the shape of the I-V curve from linear (graphene and OG-10) to nonlinear (OG-20).

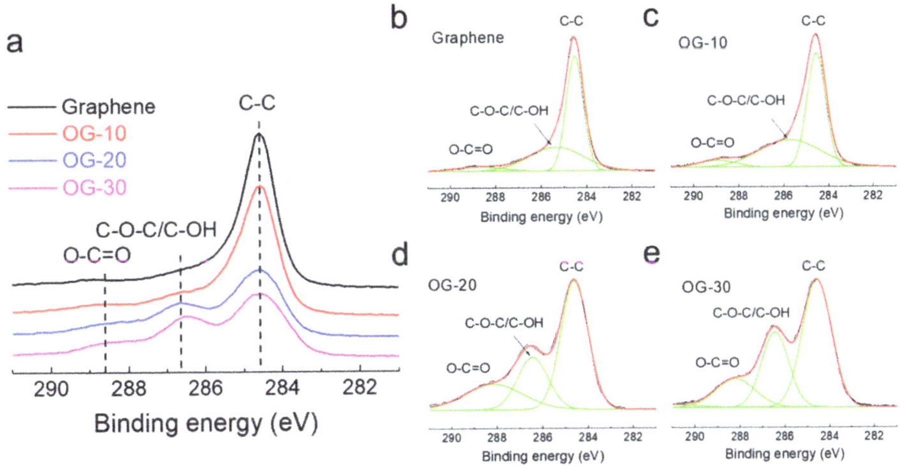

Figure 4. (**a**) XPS C 1s spectra of graphene and OG samples as a function of the H_2SO_4 concentration. Detailed peak analyses of (**b**) graphene, (**c**) OG-10, (**d**) OG-20, and (**e**) OG-30.

Table 1. Intensity ratio of the oxidized carbon and graphitic carbon (I_{OC}/I_{GC}) as a function of the H_2SO_4 concentration.

Chemical Bonds	Graphene	OG-10	OG-20	OG-30
Graphitic carbon (%)	80.38	77.06	62.59	54.33
Oxidized carbon (%)	19.62	22.94	37.41	45.67
I_{OC}/I_{GC}	0.24	0.30	0.60	0.84

Figure 5a shows the Raman spectra of graphene, OG-10, OG-20, and OG-30. The graphene spectrum shows sharp G- and 2D-bands at 1587.5 and 2688.9 cm^{-1}, with full width at half maximum (FWHM) values of 17.8 and 33.9 cm^{-1}, respectively, whereas a negligible D-band is observed at approximately 1340 cm^{-1}. In the OG-10 spectrum, D- and D′-bands (1625 cm^{-1}), which are associated with defects in graphene, appear, and the 2D-band intensity is slightly decreased compared to that of the graphene [23,24]. The defects in the OG-10 are due to vacancies, physical adsorption of oxygen-related molecules, and oxygen functional groups, including epoxide/hydroxyl, carbonyl, and carboxyl groups [22]. Compared with the OG-10, the FWHM of the G-band of OG-20 increases from 29.6 cm^{-1} (OG-10) to 63.4 cm^{-1} (OG-20) as shown in Figure 5b. Moreover, the 2D-band intensity is significantly reduced in OG-20.

The average distance between defects (L_D) in the OG can be calculated approximately using the intensity ratio of D- to G-bands (I_D/I_G) [25,26]. Ferrari and co-workers proposed a three-stage classification of the disorder of graphitic materials from graphite to amorphous carbon using Raman spectroscopy [24,27]. The first is from graphite to nanocrystalline graphite; the second is from nanocrystalline graphite to low sp^3 amorphous carbon, and the third is from low sp^3 amorphous carbon to high sp^3 amorphous carbon. To calculate the L_D of the graphene and OG specimens, we considered the following results of previous studies [28]. First, from graphene to graphene oxide, the I_D/I_G values initially increased (stage 1) and then decreased (stage 2). Second, the OG in stages 1 and 2 can have similar I_D/I_G values, while the L_D decreases. Finally, the FWHM of the G-band increases with an increase in defects; therefore, the I_D/I_G value and FWHM of the G-band must be considered simultaneously because of much larger values in stage 2. Figure 5c plots I_D/I_G versus the H_2SO_4 concentration (0% H_2SO_4 indicates graphene). The I_D/I_G values of OG specimens changed from 0.04 for graphene to 0.843 for OG-10, 1.06 for OG-20, and 0.878 for OG-30. Therefore, the I_D/I_G value increases from graphene to OG-20, but the value for OG-30 is similar to that of OG-10. In addition, the G-band FWHM of OG-20 was significantly higher than that of OG-10. Considering the above results, graphene and OG-10 are stage 1 graphitic materials, and OG-20 and OG-30 are stage 2 materials. The L_D values of graphene and OG specimens were calculated to be 49.28 (graphene), 11.04 (OG-10), 1.387 (OG-20), and 1.263 nm (OG-30), as shown in Figure 5d. These results indicate that the change in the electrical properties is due to the formation of oxygen functional groups, leading to the structural disorder.

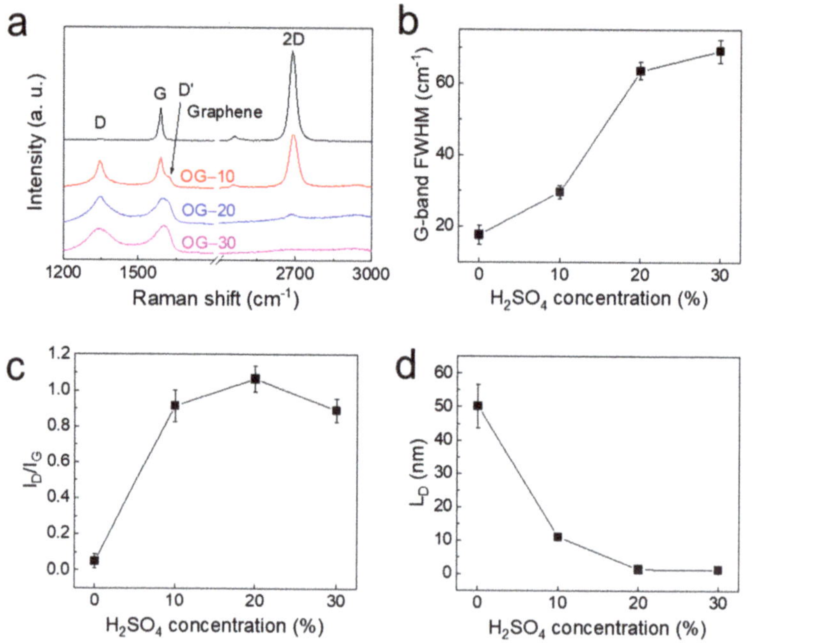

Figure 5. (**a**) Raman spectra, (**b**) G-band FWHM, (**c**) I_D/I_G, and (**d**) average distance between defects (L_D) of graphene and OGs as a function of the H_2SO_4 concentration.

The phenomenon of carrier transport in the graphene/OG-20 in-plane heterostructure can be explained using the Mott variable-range hopping (VRH) model, which involves consecutive inelastic tunneling processes between localized states [29–32]. Perhaps, OG-20 is composed of nano-sized sp^2 domains and a sp^3 matrix (disordered region), which is confirmed by the calculated L_D values as shown in Figure 5, and the disordered region plays a role as a transport barrier at the interface [32]. As shown in Figure 6, the FE-SEM image of OG-20 shows some heterogeneity in contrast, whereas the graphene appears relatively uniform. This is because of the difference in the electrical resistance between the graphene and OG-20 regions [10]. Moreover, HRTEM images (Figure 7) present the different surface morphologies of graphene and OG-20. The OG-20 has a smaller grain size than that of the graphene. The results verify the increase in the disordered regions in OG-20. The oxygen functional groups form discontinuous localized states below the conduction band, which play an important role in trapping the electrical carriers and scattering during transport. The direct tunneling of carriers through the interface between the sp^2 and sp^3 domains reveals a nonlinear I-V curve in OG-20. Therefore, it is confirmed that the electrical properties of the graphene/OG-20 in-plane heterostructures changed to a nonlinear curve with an SBH of 0.28 eV, owing to the many structural variations created by the oxygen functional groups in OG-20.

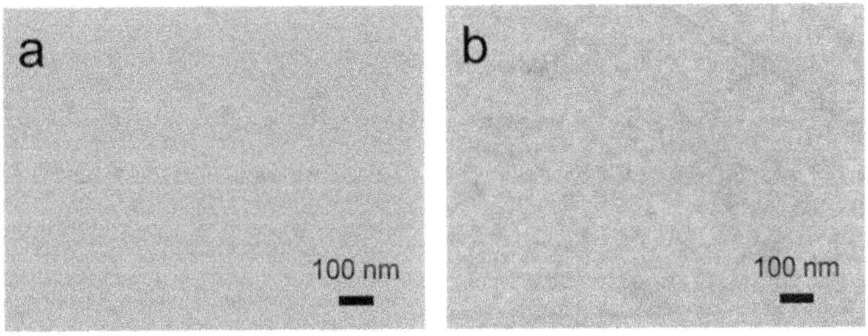

Figure 6. FE-SEM images of (**a**) graphene and (**b**) OG-20.

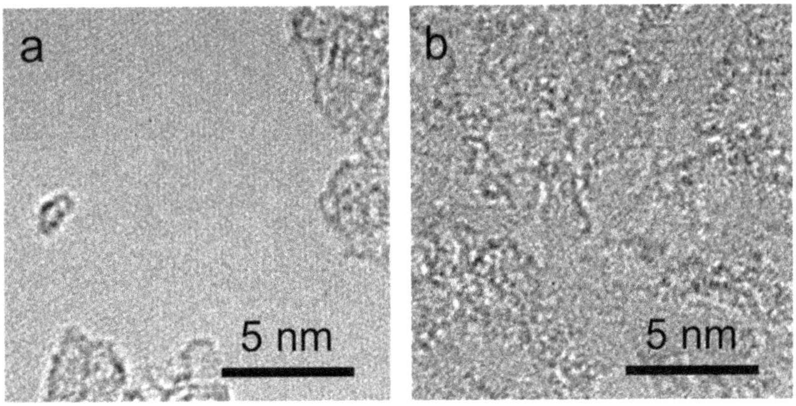

Figure 7. HRTEM images of (**a**) graphene and (**b**) OG-20.

4. Conclusions

In summary, we demonstrated variations in the electrical properties of CVD-graphene oxidized using different amounts of $KMnO_4/H_2SO_4$. The degree of oxidation of the OG specimens can be simply controlled by varying the H_2SO_4 concentration. When the H_2SO_4 concentration is above 20%, the graphene/OG in-plane heterostructure exhibits a nonlinear curve. The SBH of OG-20 was calculated to be 0.28 eV. Charge carrier transport in our devices was due to direct tunneling through interfaces between nanosized sp^2 and sp^3 regions. The sp^3 regions are formed by oxygen functional groups on the graphene surface. As a result, our simple oxidation technique enabled the modification of the electrical properties of CVD-graphene. Furthermore, the electrical characteristics of the graphene/OG in-plane heterostructures allow the design of functionalized graphene-based applications with in-plane heterostructures.

Author Contributions: Writing—original draft preparation, J.-S.C., K.-S.I. and S.-J.A.; investigation, J.-S.C., T.-K.L. and Y.-J.C.; synthesis, J.-S.C. and T.-K.L.; fabrication, J.-S.C. and K.-S.I. All authors have read and agreed to the published version of the manuscript.

Funding: This research received no external funding.

Institutional Review Board Statement: Not applicable.

Informed Consent Statement: Not applicable.

Data Availability Statement: Not applicable.

Acknowledgments: This work was supported by Kumoh National Institute of Technology.

Conflicts of Interest: The authors declare no conflict of interest.

References

1. Lonkar, S.P.; Deshmukh, Y.S.; Abdala, A.A. Recent advances in chemical modifications of graphene. *Nano Res.* **2015**, *8*, 1039–1074. [CrossRef]
2. Jin, M.; Jeong, H.-K.; Yu, W.J.; Bae, D.J.; Kang, B.R.; Lee, Y.H. Graphene oxide thin film field effect transistors without reduction. *J. Phys. D. Appl. Phys.* **2009**, *42*, 135109. [CrossRef]
3. Eda, G.; Ball, J.; Mattevi, C.; Acik, M.; Artiglia, L.; Granozzi, G.; Chabal, Y.; Anthopoulos, T.D.; Chhowalla, M. Partially oxidized graphene as a precursor to graphene. *J. Mater. Chem.* **2011**, *21*, 11217–11223. [CrossRef]
4. Jiříčková, A.; Jankovský, O.; Sofer, Z.; Sedmidubský, D. Synthesis and applications of graphene oxide. *Materials* **2022**, *15*, 920. [CrossRef] [PubMed]
5. Suhaimin, N.S.; Hanifah, M.F.R.; Azhar, M.; Jaafar, J.; Aziz, M.; Ismail, A.F.; Othman, M.H.D.; Rahman, M.A.; Aziz, F.; Yusof, N.; et al. The evolution of oxygen-functional groups of graphene oxide as a function of oxidation degree. *Mater. Chem. Phys.* **2022**, *278*, 125629. [CrossRef]
6. Bae, S.; Kim, H.; Lee, Y.; Xu, X.; Park, J.-S.; Zheng, Y.; Balakrishnan, J.; Lei, T.; Ri Kim, H.; Il Song, Y.; et al. Roll-to-roll production of 30-inch graphene films for transparent electrodes. *Nat. Nanotechnol.* **2010**, *5*, 574–578. [CrossRef] [PubMed]
7. Lee, Y.-C.; Lee, H.U.; Lee, K.; Kim, B.; Lee, S.Y.; Choi, M.-H.; Farooq, W.; Choi, J.S.; Park, J.-Y.; Lee, J.; et al. Aminoclay-conjugated TiO_2 synthesis for simultaneous harvesting and wet-disruption of oleaginous *Chlorella* sp. *Chem. Eng. J.* **2014**, *245*, 143–149. [CrossRef]
8. Zhou, H.; Yu, W.J.; Liu, L.; Cheng, R.; Chen, Y.; Huang, X.; Liu, Y.; Wang, Y.; Huang, Y.; Duan, X. Chemical vapour deposition growth of large single crystals of monolayer and bilayer graphene. *Nat. Commun.* **2013**, *4*, 2096. [CrossRef]
9. Yan, Z.; Lin, J.; Peng, Z.; Sun, Z.; Zhu, Y.; Li, L.; Xiang, C.; Samuel, E.L.; Kittrell, C.; Tour, J.M. Toward the synthesis of wafer-scale single-crystal graphene on copper foils. *ACS Nano* **2012**, *6*, 9110–9117. [CrossRef]
10. Islam, A.E.; Kim, S.S.; Rao, R.; Ngo, Y.; Jiang, J.; Nikolaev, P.; Naik, R.; Pachter, R.; Boeckl, J.; Maruyama, B. Photo-thermal oxidation of single layer graphene. *RSC Adv.* **2016**, *6*, 42545–42553. [CrossRef]
11. Yuan, J.; Ma, L.-P.; Pei, S.; Du, J.; Su, Y.; Ren, W.; Cheng, H.-M. Tuning the electrical and optical properties of graphene by ozone treatment for patterning monolithic transparent electrodes. *ACS Nano* **2013**, *7*, 4233–4241. [CrossRef]
12. Aria, A.I.; Gani, A.W.; Gharib, M. Effect of dry oxidation on the energy gap and chemical composition of CVD graphene on nickel. *Appl. Surf. Sci.* **2014**, *293*, 1–11. [CrossRef]
13. Mao, H.; Wang, R.; Zhong, J.; Zhong, S.; Chen, W. Mildly O_2 plasma treated CVD graphene as a promising platform for molecular sensing. *Carbon N. Y.* **2014**, *76*, 212–219. [CrossRef]
14. Liu, L.; Ryu, S.; Tomasik, M.R.; Stolyarova, E.; Jung, N.; Hybertsen, M.S.; Steigerwald, M.L.; Brus, L.E.; Flynn, G.W. Graphene oxidation: Thickness-dependent etching and strong chemical doping. *Nano Lett.* **2008**, *8*, 1965–1970. [CrossRef]
15. Zandiatashbar, A.; Lee, G.-H.; An, S.J.; Lee, S.; Mathew, N.; Terrones, M.; Hayashi, T.; Picu, C.R.; Hone, J.; Koratkar, N. Effect of defects on the intrinsic strength and stiffness of graphene. *Nat. Commun.* **2014**, *5*, 3186. [CrossRef] [PubMed]
16. Kim, D.C.; Jeon, D.-Y.; Chung, H.-J.; Woo, Y.; Shin, J.K.; Seo, S. The structural and electrical evolution of graphene by oxygen plasma-induced disorder. *Nanotechnology* **2009**, *20*, 375703. [CrossRef]
17. Dreyer, D.R.; Park, S.; Bielawski, C.W.; Ruoff, R.S. The chemistry of graphene oxide. *Chem. Soc. Rev.* **2010**, *39*, 228–240. [CrossRef]
18. Jia, S.; Sun, H.D.; Du, J.H.; Zhang, Z.K.; Zhang, D.D.; Ma, L.P.; Chen, J.S.; Ma, D.G.; Cheng, H.M.; Ren, W.C. Graphene oxide/graphene vertical heterostructure electrodes for highly efficient and flexible organic light emitting diodes. *Nanoscale* **2016**, *8*, 10714–10723. [CrossRef] [PubMed]
19. Wu, X.; Sprinkle, M.; Li, X.; Ming, F.; Berger, C.; de Heer, W.A. Epitaxial-graphene/graphene-oxide junction: An essential step towards epitaxial graphene electronics. *Phys. Rev. Lett.* **2008**, *101*, 26801. [CrossRef]
20. Piazza, A.; Giannazzo, F.; Buscarino, G.; Fisichella, G.; La Magna, A.; Roccaforte, F.; Cannas, M.; Gelardi, F.M.; Agnello, S. Effect of air on oxygen p-doped graphene on SiO_2. *Phys. Status Solidi* **2016**, *213*, 2341–2344. [CrossRef]
21. Yang, D.; Velamakanni, A.; Bozoklu, G.; Park, S.; Stoller, M.; Piner, R.D.; Stankovich, S.; Jung, I.; Field, D.A.; Ventrice, C.A.; et al. Chemical analysis of graphene oxide films after heat and chemical treatments by X-ray photoelectron and Micro-Raman spectroscopy. *Carbon N. Y.* **2009**, *47*, 145–152. [CrossRef]
22. Chen, J.; Zhang, Y.; Zhang, M.; Yao, B.; Li, Y.; Huang, L.; Li, C.; Shi, G. Water-enhanced oxidation of graphite to graphene oxide with controlled species of oxygenated groups. *Chem. Sci.* **2016**, *7*, 1874–1881. [CrossRef]
23. Kaniyoor, A.; Ramaprabhu, S. A Raman spectroscopic investigation of graphite oxide derived graphene. *AIP Adv.* **2012**, *2*, 32183. [CrossRef]
24. Ferrari, A.C. Raman spectroscopy of graphene and graphite: Disorder, electron–phonon coupling, doping and nonadiabatic effects. *Solid State Commun.* **2007**, *143*, 47–57. [CrossRef]
25. Tuinstra, F.; Koenig, J.L. Raman spectrum of graphite. *J. Chem. Phys.* **1970**, *53*, 1126–1130. [CrossRef]
26. Lucchese, M.M.; Stavale, F.; Ferreira, E.H.M.; Vilani, C.; Moutinho, M.V.O.; Capaz, R.B.; Achete, C.A.; Jorio, A. Quantifying ion-induced defects and Raman relaxation length in graphene. *Carbon N. Y.* **2010**, *48*, 1592–1597. [CrossRef]
27. Ferrari, A.C.; Robertson, J. Interpretation of Raman spectra of disordered and amorphous carbon. *Phys. Rev. B* **2000**, *61*, 14095–14107. [CrossRef]
28. Ferrari, A.C.; Basko, D.M. Raman spectroscopy as a versatile tool for studying the properties of graphene. *Nat. Nanotechnol.* **2013**, *8*, 235–246. [CrossRef]

29. Gómez-Navarro, C.; Weitz, R.T.; Bittner, A.M.; Scolari, M.; Mews, A.; Burghard, M.; Kern, K. Electronic transport properties of individual chemically reduced graphene oxide sheets. *Nano Lett.* **2007**, *7*, 3499–3503. [CrossRef]
30. Hill, R.M. Hopping conduction in amorphous solids. *Philos. Mag. A J. Theor. Exp. Appl. Phys.* **1971**, *24*, 1307–1325. [CrossRef]
31. Mott, N.F. Conduction in non-crystalline materials. *Philos. Mag. A J. Theor. Exp. Appl. Phys.* **1969**, *19*, 835–852. [CrossRef]
32. Sun, Y.; Kirimoto, K.; Hattori, H.; Kitamura, Y.; Fan, E.; Onishi, K. Electric field and oxygen concentration-dependent transport properties of nano-graphene oxide. *AIP Adv.* **2019**, *9*, 95010. [CrossRef]

crystals

MDPI

Article

Computational Studies of the Excitonic and Optical Properties of Armchair SWCNT and SWBNNT for Optoelectronics Applications

Yahaya Saadu Itas [1], Abdussalam Balarabe Suleiman [2], Chifu E. Ndikilar [2], Abdullahi Lawal [3], Razif Razali [4], Ismail Ibrahim Idowu [2], Mayeen Uddin Khandaker [5,6,*], Pervaiz Ahmad [7], Nissren Tamam [8], Abdelmoneim Sulieman [9] and Mohammad Rashed Iqbal Faruque [10]

1 Department of Physics, Bauchi State University Gadau, PMB 65, Gadau 751105, Nigeria; yitas@basug.edu.ng
2 Department of Physics, Federal University Dutse, Dutse 720101, Nigeria; salam@fud.edu.ng (A.B.S.); ebenechifu@yahoo.com (C.E.N.); idowu.i@fud.edu.ng (I.I.I.)
3 Department of Physics, Federal College of Education, Zaria 810282, Nigeria; labdullahi2@live.utm.my
4 Department of Physics, Faculty of Science, Universiti Teknologi Malaysia, Skudai 81310, Malaysia; razifrazali@utm.my
5 Centre for Applied Physics and Radiation Technologies, School of Engineering and Technology, Sunway University, Bandar Sunway 47500, Malaysia
6 Department of General Educational Development, Faculty of Science and Information Technology, Daffodil International University, DIU Rd, Dhaka 1341, Bangladesh
7 Department of Physics, University of Azad Jammu and Kashmir, Muzaffarabad 13100, Pakistan; pervaiz_pas@yahoo.com
8 Department of Physics, College of Sciences, Princess Nourah Bint Abdulrahman University, P.O. Box 84428, Riyadh 11671, Saudi Arabia; nmtamam@pnu.edu.sa
9 Department of Radiology and Medical Imaging, College of Applied Medical Sciences, Prince Sattam Bin Abdulaziz University, P.O. Box 422, Alkharj 11942, Saudi Arabia; a.sulieman@psau.edu.sa
10 Space Science Centre (ANGKASA), Institute of Climate Change (IPI), Universiti Kebangsaan Malaysia, Bangi 43600, Malaysia; rashed@ukm.edu.my
* Correspondence: mayeenk@sunway.edu.my

Citation: Itas, Y.S.; Suleiman, A.B.; Ndikilar, C.E.; Lawal, A.; Razali, R.; Idowu, I.I.; Khandaker, M.U.; Ahmad, P.; Tamam, N.; Sulieman, A.; et al. Computational Studies of the Excitonic and Optical Properties of Armchair SWCNT and SWBNNT for Optoelectronics Applications. *Crystals* **2022**, *12*, 870. https://doi.org/10.3390/cryst12060870

Academic Editors: Walid M. Daoush, Fawad Inam, Mostafa Ghasemi Baboli, Maha M. Khayyat and Dinadayalane Tandabany

Received: 10 May 2022
Accepted: 13 June 2022
Published: 20 June 2022

Abstract: In this study, the optical refractive constants of the (5, 5) SWBNNT and (5, 5) SWCNT systems were calculated in both parallel and perpendicular directions of the tube axis by using Quantum ESPRESSO and YAMBO code. It also extended the optical behaviors of (5, 5) SWCNT and (5, 5) SWBNNT to both perpendicular and parallel directions instead of the parallel directions reported in the literature. It also looked at the effects of the diameter of the nanotube on the optical properties instead of chiral angles. From our results, the best optical reflection was found for (5, 5) SWBNNT, while the best optical refraction was found with (5, 5) SWCNT. It was observed that the SWCNT demonstrates refraction in both parallel and perpendicular directions, while (5, 5) SWBNNT shows perfect absorption in perpendicular direction. These new features that appeared for both nanotubes in perpendicular directions were due to new optical band gaps, which appear in the perpendicular directions to both nanotubes' axis. The electron energy loss (EEL) spectrum of SWBNNT revealed the prominent π- and $\pi + \delta$- Plasmon peaks, which demonstrates themselves in the reflectivity spectrum. Furthermore, little effect of diameter was observed for the perpendicular direction to both nanotubes' axis; as such, the combined properties of (5, 5) SWBNNT and (5, 5) SWCNT materials/systems for transmitting light offer great potential for applications in mobile phone touch screens and mobile network antennas. In addition, the studies of optical properties in the perpendicular axis will help bring ultra-small nanotubes such as SWCNT and SWBNNT to the applications of next-generation nanotechnology.

Keywords: excitonic properties; SWCNT; SWBNNT; optical band gap; optical refraction; optical extinction

1. Introduction

Carbon nanotubes are tubular sheets of nanomaterials formed by rolling graphene sheets into a tubular form, and they were introduced to the field of nanoscience and nanotechnology in 1991 [1,2]. They exist in different geometries, depending on the chirality and diameter; these are armchair, zigzag, and chiral nanotubes [3]. For example, rolling the graphene sheet from its corner gives a different shape than when it is rolled from its edge. As such, the structure of materials in the form of nanotubes is determined by the chiral indices "n" and "m", which represent the coordinates of the rolling of the graphene sheet. In any case, the calculation of the chirality is based on the following Equation (1) [3]:

$$C = na_1 + ma_2 \tag{1}$$

where a_1 and a_2 are the chiral vectors of the graphene sheet. The nanotubes are called armchairs when n = m, for example (3, 3), (4, 4), (5, 5), (6, 6), and so on. It is called zigzag when 'n' is not equal to 'm', for example (5, 4), (8, 3), and so on. However, the nanotubes are called chiral when 'm' = 0. In our case, we were dealing with the armchair form of both SWCNT and SWBNNT with (5, 5) configurations. The armchair nanotubes show electrical properties similar to metals, while the zigzag and chiral nanotubes possess electrical properties similar to semiconductors. Boron nitride nanotubes are also obtained in a similar way by rolling a sheet of hexagonal boron nitride, and their chirality is also calculated based on the aforementioned Equation (1).

Since 1991, two types of carbon nanotubes have been discovered [4]. Single-walled carbon nanotubes (SWCNTs) and multi-walled carbon nanotubes (MWCNTs) have consistently been used in nanotechnology applications because of their novel thermal, mechanical, electrical, optical, and magnetic properties [5]. The size and chirality of the CNT depend on the diameter and length of the nanotubes. In today's nanoscience and technology, SWCNTs have been found to be used as sensors, LED devices, and photovoltaic and thermal reservoirs [6]. A review of the literature revealed some studies on the large armchair of SWCNTs such as (7, 7), (8, 8), (10, 10), and (12, 12) SWCNTs. Some research was also conducted on the ultra-small SWCNTs, such as (3, 3) and (4, 4) SWCNTs. However, only a few works of the literature showed the existence of (5, 5) SWCNT studies. This work was designed to solve some crucial issues regarding (5, 5) SWCNT, especially the optical properties.

On the other hand, boron nitride nanotubes (BNNTs) are rolled sheets of hexagonal boron nitride and have similar physical geometries to CNTs. They also exist in an armchair, zigzag, and chiral form [7]. They are therefore regarded as having close structural analogs of carbon nanotubes [8], because carbon atoms are alternatively replaced with boron and nitrogen atoms in a hexagonal lattice. BNNTs are perfect insulators on all configurations with a wide bandgap of 5.5 eV [9]. BNNTs show remarkable resistance to oxidation and greater thermal stability; such remarkable features make them excellent candidates in mechanical reinforcement, radiation shielding, transparent bulk composites, etc. [10]. However, both CNTs and BNNTs have some common properties [7], such as high strength, high stiffness, high thermal conductivity, low density, and high length to diameter [7].

By considering the common features found in CNTs and BNNTs, scientists have thought of combining these two novel materials in order to come up with bridged applications that can be put into next-generation technology and applications. For example, experimental studies were conducted on the synthesis of thermally stable CNT and BNNT hetero materials [11], and the results showed promising applications of CNT and BNNT in light emitting diodes (LED). Similarly, another computational approach was performed for the electronic properties of (5, 5) SWCBNNT nanostructures, which predict the excellent ability of these materials to serve as gas sensors [12]. Further review of the literature showed that many experimental and theoretical studies have been conducted on the optical properties of nanomaterials in parallel directions only (see Table 1). Besides this, very few quantum simulation studies are available on ultra-small nanotubes such as (5, 5) CBNNT systems. In order to bridge this gap, this work performed a detailed simulation on the effects of excitons across parallel and perpendicular directions to both nanotubes' axis, and such a comparative study was performed here for the first time. It also focused on the potential of two nanomaterials for comparative analysis of the optical properties of (5, 5) SWCNT and (5, 5) SWBNNT and provided some possible explanations for their use in optoelectronics. This work also explored the effects of the diameter of the nanotubes on the excitons, instead of the chiral angle, as studied by many works available in the literature (Table 1).

Table 1. Reported studies on the optical properties of nanomaterials.

S/No.	Research/Experiment	Results Obtained	Ref.
1	Optical properties of carbon nanotubes	The studies of the excitonic properties were implemented in the parallel direction to the nanotube axis. The result predicted that dark excitons affect the potential of carbon nanotubes for applications as radiation shields.	[13]
2	Optical properties of CNT single-band tight-binding approximation	The studies were applied in the parallel direction to the tube axis, and results revealed optical transitions between Van Hove singularities. The authors recommend further investigations in the perpendicular direction to the tube axis.	[14]
3	Theoretical studies on optical properties of individual carbon nanotubes	It reported that experimental and simulation studies shall be carried out on the ultra-small carbon nanotubes in all directions to confirm theoretical predictions.	[15]
4	Theoretical investigation of the optical properties of boron nitride nanotubes	The optical properties are affected by the polarization directions (parallel) as well as the radius of the nanotube.	[16]
5	Theoretical studies on vibrational and optical properties of BNNTs	Studied Raman spectroscopy and EELS of BNNTs in the parallel direction to the tube axis. It recommends that more studies need to be carried out in the perpendicular directions to bring BNNTs into the optoelectronics world.	[17]
6	Quantum simulation using quantum ESPRESSO and YAMBO codes	Studied the excitonic properties and effects of diameter on the polarization of SWCNTs and SWBNNTs in both the parallel and perpendicular directions.	This work

2. Methodology

In this research, Kohn-Sham equations were applied by implementing the DFT *ab initio* framework within the Perdiew-Burke-Emzahope (PBE) exchange functional. The energy cut-off value for the construction of the plane-wave basis set for both SWCNT and SWBNNT was achieved at 50 Ry, and the k-point value, which correlates to the ecut value, was $1 \times 1 \times 28$ k-mesh. This gave a total of 28 nk-points in the first Brillouin zone (BZ). The norm-conserving pseudopotentials were used to calculate the ion-electron's attractive interactions. Furthermore, GW-BSE calculations were performed with YAMBO code, and these were used to calculate the quasi-particle energies and optical properties of the (5, 5) SWCNT and (5, 5) SWBNNT systems. A plane-wave basis set was arranged, such that the total energy convergence was 5.5×10^{-6} Ry per carbon atom. In order to avoid inter-tube interactions, we created a vacuum from the optimized lattice parameters concerning the total energy, and the calculations were performed using a non-spin polarized DFT to save computational cost. To ensure accurate results in this study, the nanotube was appropriately relaxed to appropriate geometries. In the SWCNT, the tube length and the tube height were chosen as 5.03 Å and 3.26 Å, respectively. The chiral/translation vectors were constructed such that n = 5 and m = 5 to ensure the proper armchair chirality. The maximum force, stress, and displacements were set at 0.06 eV/Å, 0.06 GPa and 6×10^{-4} Å, respectively. The unit cell volume was 5515.67 Å^3 with lattice parameters a = 16.68 Å and c = 7.68 Å.

3. Results and Discussion

3.1. Band Gap

Computational condensed matter physics used DFT simulations to obtain the electronic and optical properties of both the bulk and nano-forms of materials with all configurations. It explained the correlated relationship between the optical bandgap and electronic bandgap of materials, as can be viewed in the ultraviolet region of the EM spectrum. As presented in Figure 1a, highly non-metallic properties can be observed from (5, 5) SWBNNT, which demonstrates a bandgap of 5.4 eV (Figure 1b); this value falls within the range of the bandgap obtained for BNNTs [18] as reported in ref. [19]. There are also more energy states in the conduction band than in the valence band for the (5, 5) SWBNNT system. The electronic bands and density of the states were also calculated for the (5, 5) SWCNT system, and zero bandgap can be seen with bands crossing the Fermi level and intersecting at the Dirac point as presented in Figure 2a. A justification is that the (5, 5) SWCNT system is metallic, as are other armchair CNTs [20]. In contrast to BNNTs, more energy states appear in the valence band than in the conduction band (Figure 2b), and the density of the state revealed three different sub-bands forming both inter-band and intra-band optical transitions (Figure 2b). There is therefore strong optical absorption as the energy of the incident photon equalizes the corresponding sub-bands [21]. In each case, there is a specific sub-band that has been identified by one integer that denotes the size of the sub-band. Three optical band gaps that are parallel to the direction of the tube axis are seen. Optical absorption occurs when the difference in sub-band energies equals the energy of the incident photon. There is a strong correlation between the chirality of the (5, 5) SWCNT and its optical transitions. This is because as the diameter of the (5, 5) SWCNT increases, there occurs a shift in the absorption peaks to greater photon energies [22].

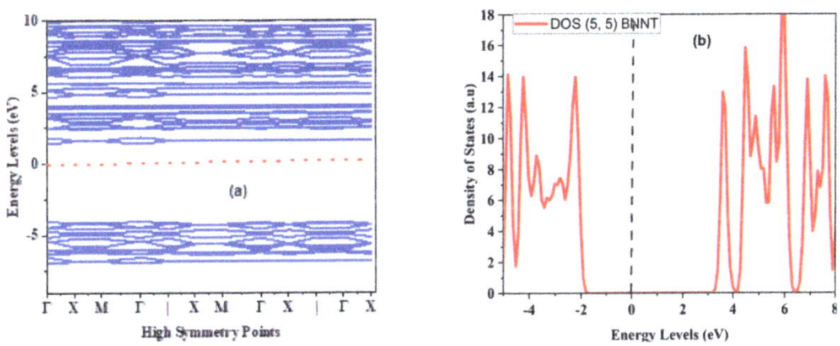

Figure 1. (**a**) The electronic bands and (**b**) density of the states of (5, 5) SWBNNTs.

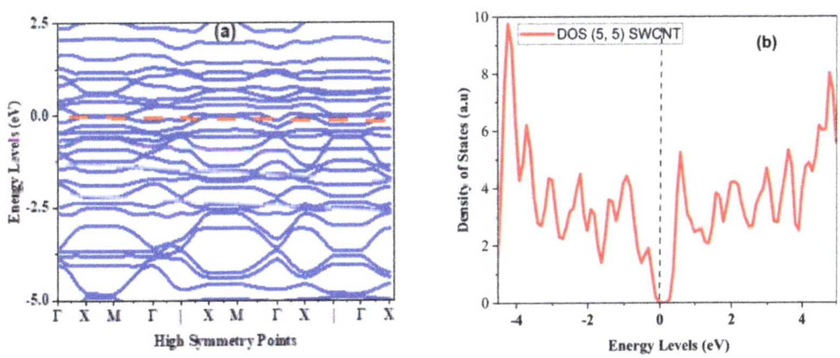

Figure 2. (**a**) The electronic bands and (**b**) density of the states of (5, 5) SWCNTs.

3.2. Dielectric Constant

Simulation analyses were conducted on the dielectric function of the real and imaginary parts of the (5, 5) SWBNNT and SWCNT in both the lower range and higher range. The studies applied to both the parallel and perpendicular directions of the nanotube axis of the two systems under study, as presented in Figure 3.

The linear response function of the (5, 5) SWBNNT and SWCNT structures to electromagnetic radiation can be explained in terms of the dielectric function [23]:

$$\varepsilon(\omega) = \varepsilon_1(\omega) + i\varepsilon_2(\omega) \tag{2}$$

where ε, is the frequency of the incident, $\varepsilon_1(\omega)$ is the frequency of the incident light for the real part, and $i\varepsilon_2(\omega)$ is the dielectric function in the imaginary part. The contribution of s and p orbitals was so affected by polarization [24], except for those orbitals close to the Fermi energy level.

$$n = \sqrt{\varepsilon_1} \tag{3}$$

The peaks of the imaginary parts of the dielectric constants of both SWCNTs and SWBNNTs illustrate the optical band gaps. In this work, the results of the optical refractive constants of the (5, 5) SWBNNT and (5, 5) SWCNT systems were found in both parallel and perpendicular directions. It was calculated that 1.3 and 1.5 eV were obtained for (5, 5) SWBNNT, and the values of 1.35 and 1.09 were obtained, respectively, for parallel and perpendicular to the tube axis for (5, 5) SWCNT. The optical gap was also larger for (5, 5) SWBNNT than (5, 5) SWCNT, which justifies that SWBNNT is a wide gap material and SWCNT is a zero-gap material. It was also observed that the nanotube diameter affected the real dielectric constant and the refractive index [25]. As such, increasing the diameter of the nanotube decreased the real dielectric constant. In condensed matter physics involving a non-spinning polarized case, we calculated the imaginary part of the dielectric function from the expression:

$$i\varepsilon_{\alpha\beta}(\omega) = \frac{\hbar^2 e^2}{\pi m^2 \omega^2} \sum_{c} \int dk \langle C_k | P^\alpha | v_k \rangle \langle v_k \left| P^\beta \right| C_k \rangle \; X \; \delta(\varepsilon_{ck} - \varepsilon_{vk} - \omega) \tag{4}$$

where m is the rest mass of an electron, e is the amount of the charge on an electron with mass m, C_k is the empty conduction band, and v_k is the occupied valence states. For calculation of the real part of the dielectric part, we adopted the Kramers–Kronig equation:

$$r\varepsilon_{\alpha\beta}(\omega) = \delta_{\alpha\beta} + \frac{2}{\pi} P \int_0^\infty \frac{\omega' i\varepsilon_{\alpha\beta}(\omega')}{\omega'^2 - \omega^2} d\omega' \tag{5}$$

where the subscripts α and β denote the components of directions.

As presented in Figure 3a, the optical gap of the (5, 5) SWCNT in the imaginary dielectric was larger in the parallel direction than in the perpendicular direction; as such, the dielectric function of (5, 5) SWCNT was anisotropic [26,27]. Furthermore, the dielectric function for the z-direction was higher than in the x-direction in the majority of the energy ranges. The presence of peaks in the imaginary part of the dielectric function indicated the occurrence of inter-band transitions [28]. In the case of the (5, 5) SWBNNT, the optical gap was larger in the perpendicular direction than in the parallel direction; three smaller peaks can be seen in the parallel direction for (5, 5) SWBNNT, and this demonstrates that BNNTs can also be anisotropic [29].

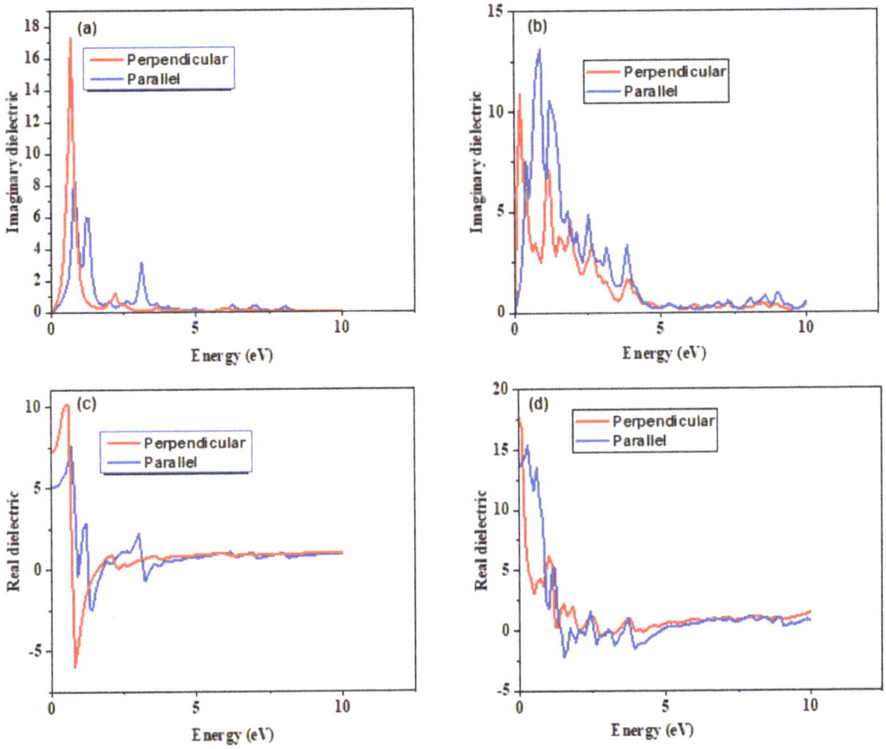

Figure 3. Dielectric functions of (5, 5) SWBNNT and SWCNT systems. (**a**) Imaginary dielectric for (5, 5) SWBNNT, (**b**) imaginary dielectric for (5, 5) SWCNT, (**c**) real dielectric for (5, 5) SWCBNNT, and (**d**) real dielectric for (5, 5) SWCNT.

3.3. Electronic Charge Distributions

The energy lost/released by an electron in traversing from one band to another was calculated for both (5, 5) SWCNT and SWBNNT systems. The calculated charge distributions of each of the two nanotubes are presented in Figure 4. In all cases, the peaks of (ω) appeared as a result of combined excitations of different photons with various frequencies. The expanded distance of the electron energy loss by the excitons in (5, 5) SWBNNT (Figure 4a) related to the absorption peak [30]. This shows that the distribution of the valence electrons has fast convergence. More energies were lost in the perpendicular direction than in the parallel direction for (5 5) SWBNNT; this can be seen by the appearance of peaks at 2.0 eV, which correspond to the energy in the plasma frequency and 2.7 eV (red) respectively. Therefore, the maximum energy was lost as a result of the effect of incident electromagnetic radiation with matter. In the (5, 5) SWCNT structure, the maximum energy loss was in the parallel direction rather than in the perpendicular direction; this can be seen from the peak at 5.0 eV presented in Figure 4b. This value is also associated with plasma energy. Furthermore, the charge distributions were more in the direction parallel to the (5, 5) SWBNNT tube axis. Sharp π-electronic peaks [31] were seen parallel and perpendicular to the tube axis, as shown in Figure 4a. Plasmon peaks were seen at 4.5 eV and 5.2 eV parallel and perpendicular to the tube axis, respectively. These results indicate that the energy loss variations had the same amplitude in (5, 5) SWCNT. The energy loss region of the EELS in Figure 4a,b is generally less than 50 eV; this is specifically called the valence electron energy loss spectroscopy (VEELS) because it is dominated by the collective excitations of the Plasmon (valence electrons) and inter-band transitions [32].

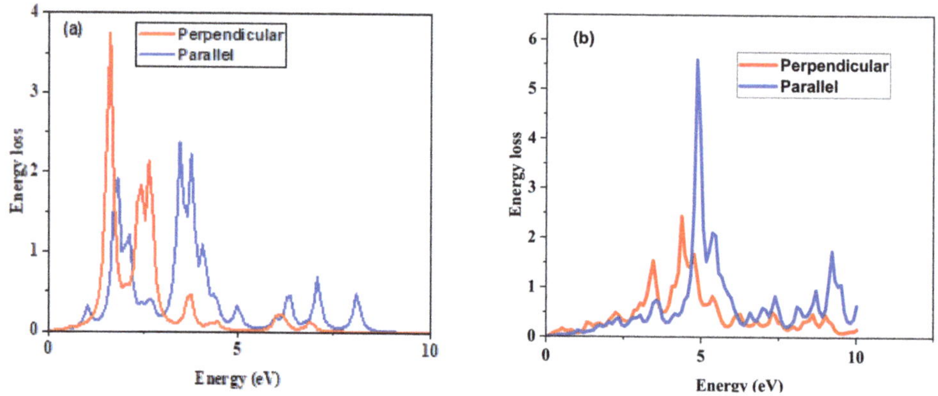

Figure 4. Electron energy loss variations of (**a**) armchair (5, 5) SWBNNT and (**b**) armchair (5, 5) SWCNT structures.

3.4. Reflection

It is very crucial to consider the volume of a monochromatic wavelength incident on a material to accurately describe its response to electromagnetic radiation. This can be achieved through studies of optical reflection and transmittance. We studied this via a spectrum of complex refractive index as a function of a single surface reflectance and phase function in both the parallel and perpendicular directions of the nanotube axis. For the (5, 5) SWBNNT structure presented in Figure 5a, the maximum reflection can be seen perpendicular to the nanotube axis, corresponding to the energy of 1.7 eV in the electromagnetic spectrum. In addition, low-intensity reflections are seen in the parallel direction of the tube axis due to π-π electronic excitations. For the (5, 5) SWCNT structure in Figure 5b, there are maximum optical reflection peaks at 0.1 eV for parallel directions and perpendicular directions. More peaks can be seen in the perpendicular direction than in the parallel direction, occurring at 5.7, 6.3, 7.7, and 9.6 eV, respectively. This illustrates that (5, 5) SWCNT has a higher optical reflection in the direction perpendicular to the nanotube axis [31].

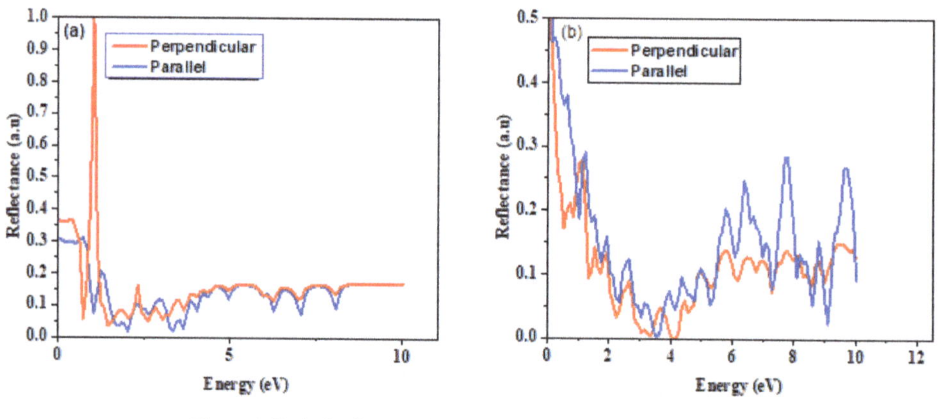

Figure 5. Optical reflection of (**a**) armchair (5, 5) SWBNNT and (**b**) armchair (5, 5) SWCNT structures.

3.5. Absorption

This study presently simulated the variation of the absorption coefficient of the (5, 5) form of SWCNT and SWBNNT structures both parallel and perpendicular to the nanotubes axis. To further reveal more about the fraction of the incident photon absorbed or transmitted, we compared the total light reflected and the total light transmitted with the incident energy. The fraction of light absorbed per given distance in a medium was termed the absorption coefficient. In investigating the optical properties of nanomaterials with zero band gaps, it was essential to calculate the optical absorption because it is another way to understand the material's optical constants, such as refractive index. We analyzed the coefficient of absorption $\alpha(\omega)$ for the incident light with frequency ω via the well-known expression $P(\omega) = \chi(\omega)E(\omega)$, with optical susceptibility $\chi(\omega)$, where $P(\omega)$ is the macroscopic polarization and $E(\omega)$ is the electric field [33]:

$$\alpha(\omega) \sim \omega Im\chi(\omega) = Im[\frac{j(\omega)}{\omega A(\omega)}] \tag{6}$$

where $j(\omega)$ is the macroscopic current density.

Figure 6 shows the region of low range and high range energies. The peaks in the low range energy appear because of the total intra-band excitations to the empty conduction states from the occupied valence states. The corresponding valence-to-conduction band transition can be seen from the peaks in the z-direction absorption spectra. The spectrum of SWBNNT presented in Figure 6a shows only one peak, which occurred at 0.7 eV in the perpendicular direction. This peak was a result of transitions between optical-electronic states [34]. Zero peaks can be seen in the parallel direction, which justifies zero absorption. The presence of a peak at 0.7 eV shows that SWBNNT only absorbed photons in the ultraviolet region, and there was no absorption in the visible region; therefore, (5, 5) SWBNNT was regarded as the poor absorber of photons in the visible region.

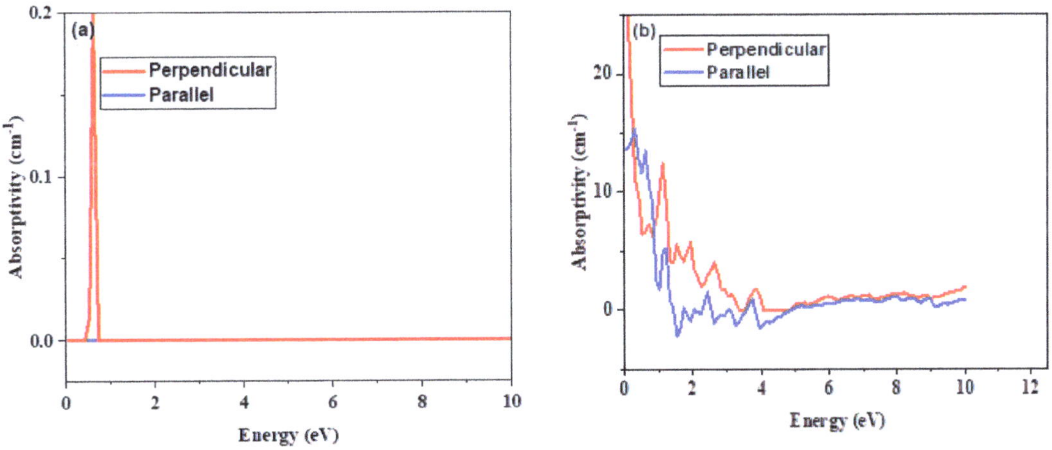

Figure 6. Optical absorptivity of (**a**) armchair (5, 5) SWBNNT and (**b**) armchair (5, 5) SWCNT structures.

In the case of (5, 5) SWCNT, peaks appeared in both parallel and perpendicular directions (Figure 6b). The peak in the parallel direction was due to the electronic transitions between sub-bands, while the peak in the perpendicular direction was due to the C-C SP^2 hybridized regions. The intensity of the peak was higher in the perpendicular direction than in the parallel direction for (5, 5) SWCNT. These two peaks demonstrate the strong absorption properties of (5, 5) SWCNT in both the UV and visible regions. To further confirm the absorption behaviors of these two nanotubes, we studied the optical properties of the (5, 5) form of both SWCNT and SWNNT via a simulation of UV-Vis [34] spectra. The results are presented in Figure 7a,b.

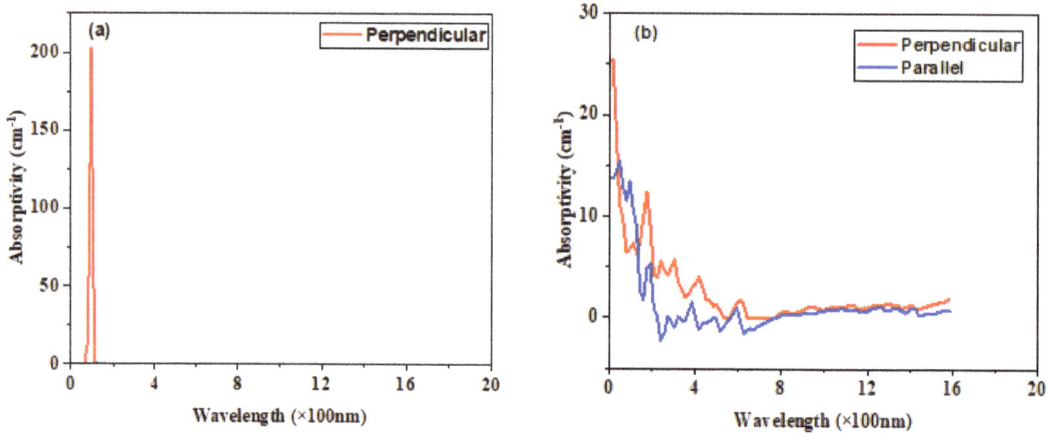

Figure 7. UV-Vis spectra of (**a**) armchair (5, 5) SWBNNT and (**b**) armchair (5, 5) SWCNT structures.

The results were calculated using Equation (7) in terms of the wavelength of monochromatic light absorbed by both systems:

$$\lambda = \frac{hC}{E} \tag{7}$$

where ***E*** is the energy of the incident photon in eV

It can be seen that armchair SWBNNT showed high absorption perpendicular to the nanotube axis only. This occurred at the wavelength range of 100 nm (see Figure 7a), which corresponded to the range of wavelength in the UV region (100–400 nm) [35]. As such, armchair SWBNNT demonstrates high absorption in the ultraviolet region. Similar analyses were carried out on armchair SWCNT, as shown in Figure 7b. In this case, (5, 5) SWCNT demonstrated high absorption in all directions in the same ultraviolet region as SWBNNT. The combined absorption properties of these nanomaterials in the ultraviolet region revealed that they can be used as impurity detectors. The similarities observed in Figure 6a,b and Figure 7a,b confirmed that electromagnetic energy is a function of wavelength.

3.6. *Refraction*

The index of refraction was fundamentally determined by the crystal structure; it is the measure of the velocity of light of a given wavelength to its velocity in a medium:

$$R(\omega) = \frac{(n-1)^2 + k^2}{(n+1)^2 + k^2} \tag{8}$$

where n is the index of refraction and k is the extinction coefficient in the real and imaginary parts, respectively, of the complex refractive index. As presented in Figure 8a, higher refractions were found parallel to the (5, 5) SWBNNT axis than the perpendicular axis; however, the extinction of refraction, which is a function of the imaginary dielectric, was higher perpendicular to the nanotube axis. Figure 8b showed that 0.1 eV was obtained for maximum optical extinction, and 2.0 eV was obtained for optical refraction, which confirms the dielectric function. Refraction was seen both parallel and perpendicular to the tube axis. The ability of (5, 5) SWCNT to refract in all directions brings it to be a potential candidate for fiber optics and magnifying lenses [35]. Our results are comparable to the results obtained for (10, 10) armchair SWCNT [36]. The coefficient of extinction is the amount of energy loss as a result of charge interactions between medium and light. We studied this to understand the number of transmissions across refracting systems, which we achieved by taking measurements of the photon frequency transmitted through SWBNNT and SWCNT as a function of energy. Figure 9a,b reveals that there was perfect transmission in all directions, and for all the SWBNNT and SWCNT structures studied, this occurred in the energy range below 1 eV.

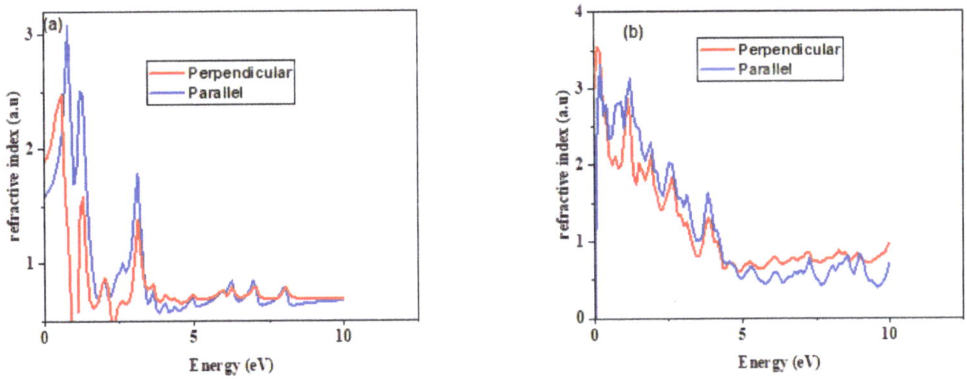

Figure 8. The optical refractive index of (**a**) armchair (5, 5) SWBNNT and (**b**) armchair (5, 5) SWCNT structures.

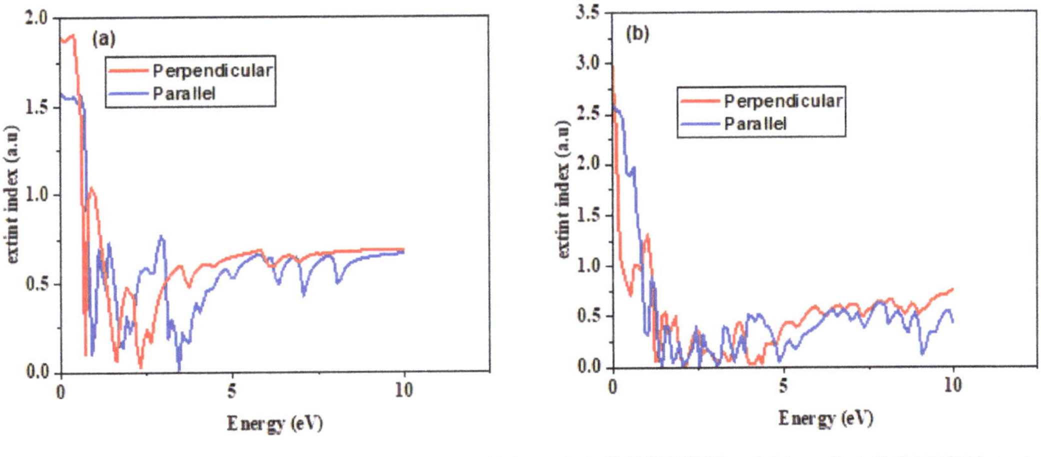

Figure 9. Optical extinction of (**a**) armchair (5, 5) SWBNNT and (**b**) armchair (5, 5) SWCNT structures.

The combined properties of (5, 5) SWBNNT and SWCNT materials to transmit light brings their applications for use in mobile phone touch screens [37] and mobile network antennas.

4. Conclusions

The excitonic properties of (5, 5) SWBNNT and (5, 5) SWCNT were studied using DFT and GW-BSE simulation methods. SWBNNT demonstrated optical bandgaps in both the parallel and perpendicular directions, while SWCNT showed a weaker optical gap in the perpendicular direction. It is generally reported that the electron energy is lost due to the valence electron energy loss spectroscopy because it is dominated by the collective excitations of the Plasmon (valence electrons) and inter-band transitions. However, in this study, the calculated charge distribution revealed that electron energy was lost due to the combined excitations of different photons with various frequencies. In addition to their various optical responses in various directions, SWBNNT showed reflection only in the perpendicular direction, while SWCNT reflected in all directions. The (5, 5) SWBNNT absorbed phonon only in the perpendicular direction, which suggests that it cannot be used for perfect absorption applications. On the other hand, (5, 5) SWCNT showed peaks in both directions; the peak in the parallel direction was due to the electronic transitions between sub-bands, while the peak in the perpendicular direction was due to the C-C SP^2 hybridized regions. It is therefore regarded that SWCNT possesses strong absorption properties and is suitable to be used in absorption applications, such as impurity detectors. The dielectric property of (5, 5) SWCNT confirmed that 0.1 eV was obtained for maximum optical extinction and 2.0 eV was obtained for optical refraction. This showed that SWCNT transmits and refracts light, and refractions were also seen in both parallel and perpendicular directions, which brings its application to fiber optics and magnifying lenses. There was also a perfect transmission at low energy ranges for both systems. The combined properties of (5, 5) SWBNNT and SWCNT materials to transmit light make them suitable to be used in mobile phone touch screens and mobile network antennas.

Author Contributions: Conceptualization, Y.S.I. and A.B.S.; methodology, Y.S.I., I.I.I., and C.E.N.; software, Y.S.I., C.E.N., R.R., M.R.I.F., and A.L.; formal analysis, Y.S.I., I.I.I., and A.B.S.; resources, A.S. and N.T.; data curation, P.A. and Y.S.I.; writing—original draft preparation, Y.S.I.; writing—review and editing, M.U.K.; visualization, P.A., M.R.I.F. and R.R.; funding acquisition, A.S. and N.T. All authors have read and agreed to the published version of the manuscript.

Funding: This work was supported by the Princess Nourah bint Abdulrahman University Researchers Supporting Project (Grant No. PNURSP2022R12), Princess Nourah bint Abdulrahman University, Riyadh, Saudi Arabia.

Institutional Review Board Statement: Not applicable.

Informed Consent Statement: Not applicable.

Data Availability Statement: No data reported.

Acknowledgments: The authors express their gratitude to Princess Nourah bint Abdulrahman University Researchers Supporting Project (Grant No. PNURSP2022R12), Princess Nourah bint Abdulrahman University, Riyadh, Saudi Arabia. The Authors also acknowledged Federal University Dutse-Nigeria, for giving resource training to the lead researcher, Bauchi State University Gadau-Nigeria, and the Tertiary Education Trust Fund (TETFund)-Nigeria for providing resource, funds and avenue for the successful conduct of this research.

Conflicts of Interest: The authors declare no conflict of interest.

References

1. Saifuddin, N.; Raziah, A.Z.; Junizah, A.R. Carbon Nanotubes: A Review on Structure and Their Interaction with Proteins. *J. Chem.* **2012**, *2013*, 1–18. [CrossRef]
2. Itas, Y.S.; Ndikilar, C.E.; Zangina, T. Carbon Nanotubes: A Review of Synthesis and Characterization Methods/Techniques. *Int. J. Sci. Technoledge* **2020**, *8*. [CrossRef]
3. Eatemadi, A.; Daraee, H.; Karimkhanloo, H.; Kouhi, M.; Zarghami, N.; Akbarzadeh, A.; Abasi, M.; Hanifehpour, Y.; Joo, S.W. Carbon nanotubes: Properties, synthesis, purification, and medical applications. *Nanoscale Res. Lett.* **2014**, *9*, 393. [CrossRef] [PubMed]
4. Pandey, P.; Dahiya, M. Carbon nanotubes: Types, methods of preparation and applications. *Int. J. Pharm. Sci. Res.* **2016**, *1*, 15–21.
5. Dresselhaus, M.S.; Dresselhaus, G.; Charlier, J.-C.; Hernandez, E.R. Electronic, thermal and mechanical properties of carbon nanotubes. *Philos. Trans. R. Soc. London. Ser. A Math. Phys. Eng. Sci.* **2004**, *362*, 2065–2098. [CrossRef]
6. Abdalla, S.; AlMarzouki, F.; Al-Ghamdi, A.A.; Abdel-Daiem, A. Different Technical Applications of Carbon Nanotubes. *Nanoscale Res. Lett.* **2015**, *10*, 1–10. [CrossRef]
7. Celik-Aktas, A.; Zuo, J.M.; Stubbins, J.F.; Tang, C.; Bando, Y. Structure and chirality distribution of multiwalled boron nitride nanotubes. *Appl. Phys. Lett.* **2005**, *86*, 133110. [CrossRef]
8. Dolati, S.; Fereidoon, A.; Kashyzadeh, K.R. A Comparison Study between Boron nitride nanotubes and Carbon nanotubes. *Int. J. Emerg. Technol. Adv. Eng.* **2012**, *2*, 1–11.
9. Saikia, N.; Pati, S.K.; Deka, R.C. First principles calculation on the structure and electronic properties of BNNTs functionalized with isoniazid drug molecule. *Appl. Nanosci.* **2012**, *2*, 389–400. [CrossRef]

10. Zhang, D.; Zhang, S.; Yapici, N.; Oakley, R.; Sharma, S.; Parashar, V.; Yap, Y.K. Emerging Applications of Boron Nitride Nanotubes in Energy Harvesting, Electronics, and Biomedicine. *ACS Omega* **2021**, *6*, 20722–20728. [CrossRef]
11. Itas, Y.S.; Ndikilar, C.E.; Zangina, T.; Hafeez, H.Y.; Safana, A.A.; Khandaker, M.U.; Ahmad, P.; Abdullahi, I.; Olawumi, B.K.; Babaji, M.A.; et al. Synthesis of Thermally Stable *h*-BN-CNT Hetero-Structures via Microwave Heating of Ethylene under Nickel, Iron, and Silver Catalysts. *Crystals* **2021**, *11*, 1097. [CrossRef]
12. Itas, Y.S.; Suleiman, A.B.; Ndikilar, C.E.; Lawal, A.; Razali, R.; Khandaker, M.U.; Ahmad, P.; Tamam, N.; Sulieman, A. The Exchange-Correlation Effects on the Electronic Bands of Hybrid Armchair Single-Walled Carbon Boron Nitride Nanostructure. *Crystals* **2022**, *12*, 394. [CrossRef]
13. Saito, R.; Dresselhaus, M.S. Optical Properties of Carbon Nanotubes. In *Carbon Nanotubes and Graphene*; Elsevier: Amsterdam, The Netherlands, 2014; Volume 2014, pp. 77–98. ISBN 9780080982328. [CrossRef]
14. Pacheco, M.; Barticevic, Z.; Latge, A.; Rocha, C.G. Optical properties of carbon nanotubes under external electric fields. *Braz. J. Phys.* **2006**, *36*, 440–442. [CrossRef]
15. Blancon, J.-C. Optical absorption and electronic properties of individual carbon nanotubes. Ph.D. Thesis, Université Claude Bernard—Lyon I, Villeurbanne, France, 2013.
16. Chen, R.B.; Shyu, F.L.; Chang, C.P.; Lin, M.F. Optical Properties of Boron Nitride Nanotubes. *J. Phys. Soc. Jpn.* **2002**, *17*, 2286–2289. [CrossRef]
17. Wirtz, L.; Rubio, A. Optical and Vibrational Properties of Boron Nitride Nanotubes. *Irorun* **2009**, *13*, 105–148. [CrossRef]
18. Li, F.; Lu, J.; Tan, G.; Ma, M.; Wang, X.; Zhu, H. Boron nitride nanotubes composed of four- and eight-membered rings. *Phys. Lett. A* **2019**, *383*, 76–82. [CrossRef]
19. Kweitsu, E.O.; Armoo, S.K.; Kan-Dapaah, K.; Abavare, E.K.K.; Dodoo-Arhin, D.; Yaya, A. Comparative Study of Phosgene Gas Sensing Using Carbon and Boron Nitride Nanomaterials—A DFT Approach. *Molecules* **2020**, *26*, 120. [CrossRef]
20. Tang, Y.; Lu, J.; Liu, D.; Yan, X.; Yao, C.; Zhu, H. Structural Derivative and Electronic Property of Armchair Carbon Nanotubes from Carbon Clusters. *J. Nanomater.* **2017**, *2017*, 1–11. [CrossRef]
21. Piazza, V.; Gemmi, M. *Optical properties of boron nitride nanotubes: Potential exploitation in nanomedicine*; William Andrew Publishing: Norwich, NY, USA, 2016; pp. 139–147. [CrossRef]
22. Xu, Z.; Zhang, W.; Zhu, Z.; Ren, C.; Li, Y.; Huai, P. Effects of tube diameter and chirality on the stability of single-walled carbon nanotubes under ion irradiation. *J. Appl. Phys.* **2009**, *106*, 43501. [CrossRef]
23. Jafarzadeh, N.; Nadafan, M.; Malekfar, R.; Shakeri-Zadeh, A.; Meidanchi, A.; Eynali, S. Structural, optical and dielectric studies of Ag nanoparticles decorated by herceptin. *Phys. E Low-dimensional Syst. Nanostructures* **2019**, *114*, 113562. [CrossRef]
24. Nicolás, G.; Dorantes-Dávila, J.; Pastor, G. Calculation of orbital polarization effects in small Co clusters. *Comput. Mater. Sci.* **2005**, *35*, 292–296. [CrossRef]
25. Guo, X.; Yu, D.; Wu, J.; Min, C.; Guo, R. Effects of nanotube modification on the dielectric behaviors and mechanical properties of multiwall carbon nanotubes/epoxy composites. *Polym. Eng. Sci.* **2012**, *53*, 370–377. [CrossRef]
26. Slepchenkov, M.M.; Shmygin, D.S.; Zhang, G.; Glukhova, O.E. Controlling anisotropic electrical conductivity in porous graphene-nanotube thin films. *Carbon* **2020**, *165*, 139–149. [CrossRef]
27. Beni, Y.T.; Mehralian, F.; Zeverdejani, M.K. Free vibration of anisotropic single-walled carbon nanotube based on couple stress theory for different chirality. *J. Low Freq. Noise Vib. Act. Control* **2017**, *36*, 277–293. [CrossRef]
28. Wang, H.; Tam, F.; Grady, N.K.; Halas, N.J. Cu Nanoshells: Effects of Interband Transitions on the Nanoparticle Plasmon Resonance. *J. Phys. Chem. B* **2005**, *109*, 18218–18222. [CrossRef]
29. Zhan, Y.; Lago, E.; Santillo, C.; Castillo, A.E.D.R.; Hao, S.; Buonocore, G.G.; Chen, Z.; Xia, H.; Lavorgna, M.; Bonaccorso, F. An anisotropic layer-by-layer carbon nanotube/boron nitride/rubber composite and its application in electromagnetic shielding. *Nanoscale* **2020**, *12*, 7782–7791. [CrossRef]
30. Chen, H.; Chen, Y.; Liu, Y.; Xu, C.-N.; Williams, J.S. Optical properties of BN nanotubes. In Proceedings of the 2006 International Conference on Nanoscience and Nanotechnology, Brisbane, Australia, 3–7 July 2006. [CrossRef]
31. Gharbavi, K.; Badehian, H. Optical properties of armchair (7, 7) single walled carbon nanotubes. *AIP Adv.* **2015**, *5*, 77155. [CrossRef]
32. Toudert, J.; Serna, R. Interband transitions in semi-metals, semiconductors, and topological insulators: A new driving force for plasmonics and nanophotonics [Invited]. *Opt. Mater. Express* **2017**, *7*, 2299–2325. [CrossRef]
33. Radzwan, A.; Lawal, A.; Shaari, A.; Chiromawa, I.M.; Ahams, S.T.; Ahmed, R. First-principles calculations of structural, electronic, and optical properties for Ni-doped Sb2S3. *Comput. Condens. Matter* **2020**, *24*, e00477. [CrossRef]
34. Bolotin, D.S.; Novikov, A.S.; Kolesnikov, I.E.; Suslonov, V.V.; Novozhilov, Y.; Konzhina, O.; Dorogov, M.; Krasavin, M.; Kukushkin, V.Y. Phosphorescent Platinum(II) Complexes Featuring Chelated Acetoxime Pyrazoles: Synthetic, Structural, and Photophysical Study. *ChemistrySelect* **2016**, *1*, 456–461. [CrossRef]
35. Tan, Y.C. Chemical Sensing Applications of Carbon Nanotube-Deposited Optical Fibre Sensors. *Chemosensors* **2018**, *6*, 55. [CrossRef]
36. Fahad, J.; Bhardwaj, A.K. Preparing Carbon Nanotubes (Cnts) for Optical System Applications. *Int. J. Nanotechnol. Appl.* **2013**, *2*, 21–38.
37. Jiang, L.; Liu, P.; Liu, C.; Fan, S. Enhanced light transmission of carbon nanotube film by ultrathin oxide coatings. *AIP Adv.* **2020**, *10*, 75304. [CrossRef]

crystals

MDPI

Article

Fullerene Movement in a Carbon Nanocontainer

Alexey M. Bubenchikov [1], Mikhail A. Bubenchikov [1], Anna S. Chelnokova [1,*], Dmitriy V. Mamontov [1] and Alexander V. Lun-Fu [2]

[1] Regional Scientific and Educational Mathematical Center, National Research Tomsk State University, 36 Lenin Ave., 634050 Tomsk, Russia
[2] LLC "Gazprom Transgaz Tomsk", 9 Frunze St., 634029 Tomsk, Russia
* Correspondence: smolina-nyuta@mail.ru

Abstract: The character of C_{60} fullerene motion inside a cylindrical carbon nanocontainer with flat graphene covers has been studied. A new trajectory approach to describe the three-dimensional motion of fullerene is presented, based on the use of motion equations for its center of mass in the field of van der Waals forces, as well as the rotational motion equations written in the absolute basis. A high-precision computational technology for implementing this approach is described. Calculations have shown the components of the C_{60} angular velocity change stepwise. This character of the change in rotation is determined by the impacts of the fullerene on the wall of the container. As a result of these impacts, the energy of rotation changes abruptly.

Keywords: fullerene; carbon nanocontainer; molecular dynamics; trajectory approach

Citation: Bubenchikov, A.M.; Bubenchikov, M.A.; Chelnokova, A.S.; Mamontov, D.V.; Lun-Fu, A.V. Fullerene Movement in a Carbon Nanocontainer. *Crystals* **2022**, *12*, 1179. https://doi.org/10.3390/cryst12081179

Academic Editors: Walid M. Daoush, Fawad Inam, Mostafa Ghasemi Baboli and Maha M. Khayyat

Received: 1 June 2022
Accepted: 19 August 2022
Published: 22 August 2022

1. Introduction

Molecular structures composed of fullerenes, graphenes and nanotubes have great potential applications in nanotechnology. They can be used as power converters, pumps, water desalination and cooling devices, drug delivery containers, storage shuttles, switches, sensors and nano-oscillators.

Encapsulation of an ion [1] or a water molecule [2] into a fullerene makes it possible to obtain such a tool in controlling the movement of fullerene as a free charge [3] or a dipole moment [4]. This opens up great opportunities for the development of new nanosized devices. Articles, such as [5–7], talk about new applications in water desalination and cooling technologies, energy conversion and targeted drug delivery. The authors of [8] also propose a fully controlled electromechanical device for pumping liquid at the nanoscale. In [9], the mechanics of an oscillator in the form of a fullerene located inside a hexagonal packing of a carbon nanotube beam were considered. The authors of [10] presented a C_{60} oscillator placed in a groove of a graphene nanoribbon. In [11], the interaction of fullerene with a carbon nanotube was studied. The results were obtained both experimentally and using molecular dynamics calculations. In [12], using the methods of molecular dynamics, results were obtained on the dynamic state of buckyballs enclosed in a carbon nanotube closed at both ends. In [13], the bending strength of a nanotube containing fullerenes was analyzed. In [14], based on the theory based on Green's function method, the properties of charge transfer through the $CNT/C_{60}/CNT$ system were studied. The authors of [15] added another $Li/C_{60}/Li$ to the considered system and showed that these systems radically differ from each other in their physical properties. The articles [16,17] present the results of laboratory research and synthesis of nanopods with C_{60}, C_{70} and C_{82} fullerenes. In [18], the dynamic state of fullerenes in similar carbon pods was calculated. However, only the vibrations of fullerenes, that is, the displacements of their centers of mass, were determined.

In this paper, we propose a new way to describe the three-dimensional motion of a buckyball based on the use of motion equations for its center of mass, as well as the rotational motion equations written in the absolute basis. A high-precision algorithm for the numerical implementation of the developed approach is also proposed. On this basis, the dynamic state of the buckyball is studied, located in a cylindrical carbon container (Figure 1), 3.2 nm in diameter, 4 nm high, with flat graphene covers.

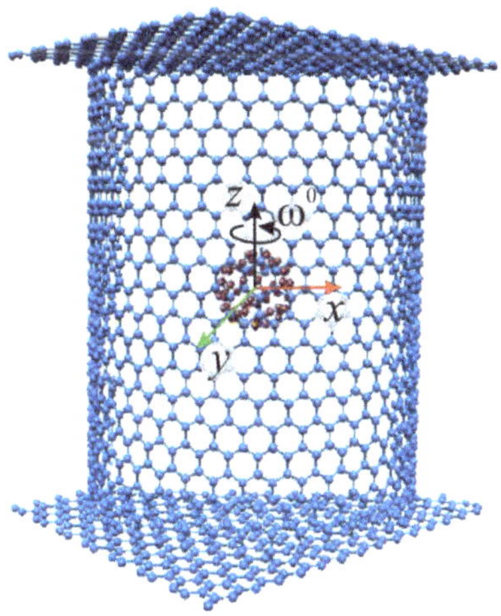

Figure 1. Carbon nanocontainer with C_{60} fullerene. x,y,z-coordinates of the absolute basis, ω-the angular direction.

2. Materials and Methods

The moving fullerene molecule in our consideration is rigid, i.e., not subject to elastic deformation. In addition, vibrations of carbon atoms in a buckyball are not taken into account. Since the C_{60} molecule is a fairly large object compared to the volume of a nanocontainer, the processes of conversion of rotational energy into translational energy will play an important role in a detailed description of its movements. The rotation of a supermolecule can be effectively described in terms of the atom–atom interaction approach. The application of this approach reduces the problem of the interaction of supermolecules to the problem of the interaction of a set of force centers, the role of which, in relation to the molecular structures under consideration, is played by carbon atoms. Thus, interactions of each C_{60} atom with all atoms in the fullerene environment will be included in the calculation. Within the framework of the considered approach, fullerene rotations around its own center of mass can be determined on the basis of the theorem on the change in the angular momentum of a rotating object. The left side of the notation of this theorem includes the time derivative of the angular momentum and the right side is the sum of the moments of forces acting on the force centers of the fullerene.

When considering the motion of immutable molecular structures, representative points are naturally identified. Such points are the atoms that make up the structures under consideration. In this case, the position of the structure in space is completely determined by the coordinates of these points. Thus, if we include the coordinates of atoms in the number of unknown functions and find kinematic relations to determine these coordinates, then we can build a solution to the problem of the position of a molecular structure in space in an absolute basis, i.e., without Euler angles. The mentioned kinematic relations in essence express the theorem of addition of velocities for the complex motion of representative points.

If in a macroscopic body, we introduce into consideration three representative points that do not lie on one straight line, then this method can be used to determine the position in space of the so-called "solid" bodies. However, in this case it is necessary to tabulate the instantaneous values of all components of the inertia tensor. This can be done in advance before solving the problem of determining the orientation of the body in space.

Fullerene rotation is described by equations for the angular momentum projections. These equations in the absolute basis have the form:

$$\frac{dK_x}{dt} = \sum_{k=1}^{N_1}((y_k - y_c)Z_k - (z_k - z_c)Y_k) = L_x, \tag{1}$$

$$\frac{dK_y}{dt} = \sum_{k=1}^{N_1}((z_k - z_c)X_k - (x_k - x_c)Z_k) = L_y, \tag{2}$$

$$\frac{dK_z}{dt} = \sum_{k=1}^{N_1}((x_k - x_c)Y_k - (y_k - y_c)X_k) = L_z. \tag{3}$$

Here x_c, y_c, z_c are the coordinates for the fullerene center of mass; x_k, y_k, z_k—coordinates of carbon atoms in fullerene; N_1 is the number of atoms in fullerene; K_x, K_y, K_z are the projections of the angular momentum.

Equations (1)–(3) are considered as equations for determining the angular velocity projections of the molecular structure in the absolute basis. Since these quantities enter under the sign of the differential in (1)–(3), the initial conditions are needed for the angular velocity components:

$$t = 0,\ \omega_x = \omega_x^0,\ \omega_y = \omega_y^0,\ \omega_z = \omega_z^0. \tag{4}$$

The resulting projections of the forces acting on each individual fullerene atom are determined by the following sums:

$$X_k = -\sum_{j=1}^{N_2} \frac{\partial}{\partial x} U\left(\rho_{jk}\right),\ Y_k = -\sum_{j=1}^{N_2} \frac{\partial}{\partial y} U\left(\rho_{jk}\right),\ Z_k = -\sum_{j=1}^{N_2} \frac{\partial}{\partial z} U\left(\rho_{jk}\right), \tag{5}$$

where N_2 is the number atoms in nanocontainer; $U\left(\rho_{jk}\right)$—interaction potential of the kth fullerene atom and the jth nanocontainer atom; ρ_{jk}—distances between atoms. In the present calculations, we take it in the form of classical *LJ*-potential with a reduced value of the characteristic interaction energy, since both carbon atoms are in the bound state in molecular structures, i.e., are not free. Nevertheless, this potential, like any other interaction potential of two neutral atoms applied to molecular fragments, makes it possible to isolate potential wells near the surfaces of the nanocontainer. The resulting potential wells make the potential field of interactions inhomogeneous.

It is easy to show that the projections of the angular momentum are defined as the result of the product of a string representing the vector of instantaneous angular velocity and the inertia tensor of a rotating object:

$$K_x = A\omega_x + F\omega_y + E\omega_z, \tag{6}$$

$$K_y = F\omega_x + B\omega_y + D\omega_z, \tag{7}$$

$$K_z = E\omega_x + D\omega_y + C\omega_z. \tag{8}$$

In this case, the components in the inertia tensor are determined by the following formulas:

$$\begin{gathered} A = m\sum\left((y_i - y_c)^2 + (z_i - z_c)^2\right),\ B = m\sum\left((z_i - z_c)^2 + (x_i - x_c)^2\right), \\ C = m\sum\left((x_i - x_c)^2 + (y_i - y_c)^2\right),\ D = -m\sum(y_i - y_c)(z_i - z_c)_i, \\ E = -m\sum(z_i - z_c)(x_i - x_c),\ F = -m\sum(x_i - x_c)(y_i - y_c). \end{gathered} \tag{9}$$

Here m is the mass of a carbon atom.

The atoms that make up the fullerene participate in a complex movement, namely, together with the center of mass of the fullerene and in rotations around it. Since the velocity addition theorem is valid for a complex motion, the corresponding additive formulas will be valid for the coordinates of the points involved in the complex motion. Therefore, the coordinates of carbon atoms in a molecule fullerene are determined as follows:

$$\begin{aligned}\frac{dx_k}{dt} &= u_c + \omega_y(z_k - z_c) - \omega_z(y_k - y_c),\\ \frac{dy_k}{dt} &= v_c + \omega_z(x_k - x_c) - \omega_x(z_k - z_c),\\ \frac{dz_k}{dt} &= w_c + \omega_x(y_k - y_c) - \omega_y(x_k - x_c).\end{aligned} \tag{10}$$

These differential equations are integrated under the following initial conditions:

$$t = 0,\ x_k = x_k^0,\ y_k = y_k^0,\ z_k = z_k^0\ (k = \overline{1, N_1}). \tag{11}$$

Index zero at the top means the initial values of the coordinates.

The coordinates and velocities of the moving center of mass of the molecule obey the following differential equations:

$$\frac{dx_c}{dt} = u_c,\ \frac{dy_c}{dt} = v_c,\ \frac{dz_c}{dt} = w_c, \tag{12}$$

$$\frac{du_c}{dt} = \frac{1}{M}\sum_{k=1}^{N_1} X_k = U_c,\ \frac{dv_c}{dt} = \frac{1}{M}\sum_{k=1}^{N_1} Y_k = V_c,\ \frac{dw_c}{dt} = \frac{1}{M}\sum_{k=1}^{N_1} Z_k = W_c. \tag{13}$$

Here M is the fullerene mass.

$$\begin{aligned}t = 0,\ x_c = x_c^0,\ y_c = y_c^0,\ z_c = z_c^0,\\ u_c = u_c^0,\ v_c = v_c^0,\ w_c = w_c^0.\end{aligned} \tag{14}$$

3. Calculation Scheme

Let $x^n,\ y^n,\ z^n$ be the Cartesian coordinates of the atoms of the molecule on the nth time layer. Here and below, for reasons of notational simplicity, we will omit the subscript "k" of the quantities under consideration. The angular velocity components $\omega_x^n,\ \omega_y^n,\ \omega_z^n$ will also refer to the nth time layer. In particular, for $n = 0$, these are the initial values of the required quantities. Since we use a scheme that formally has the fourth order of computational accuracy with respect to the time step, its implementation requires the calculation of the desired values at four positions of each individual time step. Thus, in the following notation, $x_1,\ y_1,\ z_1;\ x_2,\ y_2,\ z_2;\ x_3,\ y_3,\ z_3;\ x_4,\ y_4,\ z_4$ are coordinates in four intermediate positions. Here, the first index corresponding to the number of carbon atoms in the fullerene is deliberately omitted (for ease of notation). Obviously, the values in the first position should be:

$$\begin{aligned}&x_1 = x^n,\ y_1 = y^n,\ z_1 = z^n;\ \omega_{x1} = \omega_x^n,\ \omega_{y1} = \omega_y^n,\ \omega_{z1} = \omega_z^n;\\ &u_{c1} = u_c^n,\ v_{c1} = v_c^n,\ w_{c1} = w_c^n.\end{aligned}$$

These values, as well as formulas (6)–(9), calculate $A,\ B,\ C,\ D,\ E,\ F,\ L_x,\ L_y,\ L_z,\ K_x,\ K_y,\ K_z$ in this position.

Before finding the coordinate values in the second position of the time step, it should be remembered that for the complex movement of a node in a molecular structure, the velocity addition theorem is valid:

$$\frac{dx}{dt} = u_c + u_r,\ \frac{dy}{dt} = v_c + v_r,\ \frac{dz}{dt} = w_c + w_r. \tag{15}$$

Here $u_r,\ v_r,\ w_r$ are the atomic velocities arising due to the rotation of the molecule about its center of mass. Taking into account (15), we calculate the coordinates of the nodes in the second position:

$$x_2 = x^n + \frac{\Delta t}{2}(u_{c1} + u_{r1}),\ y_2 = y^n + \frac{\Delta t}{2}(v_{c1} + v_{r1}),\ z_2 = z^n + \frac{\Delta t}{2}(w_{c1} + w_{r1}). \tag{16}$$

The projections of the angular momentum in this position will be as follows:

$$K_{x2} = K_x^n + \frac{\Delta t}{2} L_{x1},\ K_{y2} = K_y^n + \frac{\Delta t}{2} L_{y1}, K_{z2} = K_z^n + \frac{\Delta t}{2} L_{z1}. \quad (17)$$

Substituting into the left parts of the last relations of representation (6)–(8), we obtain the values of the projections of the angular momentum in the second position:

$$\left.\begin{aligned} K_{x2} &= A_2\omega_{x2} + F_2\omega_{y2} + E_2\omega_{z2}, \\ K_{y2} &= F_2\omega_{x2} + B_2\omega_{y2} + D_2\omega_{z2}, \\ K_{z2} &= E_2\omega_{x2} + D_2\omega_{y2} + C_2\omega_{z2}, \end{aligned}\right\} \quad (18)$$

We obtain a system of three linear algebraic equations for ω_{x2}, ω_{y2}, ω_{z2} values of fullerene angular velocity projections in the second position. Now, using relations (13), we find:

$$u_{c2} = u_c^n + \frac{\Delta t}{2} U_{c1},\ v_{c2} = v_c^n + \frac{\Delta t}{2} V_{c1},\ w_{c2} = w_c^n + \frac{\Delta t}{2} W_{c1}.$$

When the projections of the angular velocities in the second position are found and with the help of (13), the projections of the velocities of the center of mass in the same position are found, it is possible to determine the coordinates of the nodes in the molecular structure already in the third position:

$$x_3 = x^n + \frac{\Delta t}{2}(u_{c2} + u_{r2}),\ y_3 = y^n + \frac{\Delta t}{2}(v_{c2} + v_{r2}),\ z_3 = z^n + \frac{\Delta t}{2}(w_{c2} + w_{r2}).$$

and

$$\begin{aligned} u_{r2} &= (z_2 - z_{c2})\omega_{y2} - (y_2 - y_{c2})\omega_{z2}, \\ v_{r2} &= (x_2 - x_{c2})\omega_{z2} - (z_2 - z_{c2})\omega_{x2}, \\ w_{r2} &= (y_2 - y_{c2})\omega_{x2} - (x_2 - x_{c2})\omega_{y2}. \end{aligned}$$

Here u_{r2}, v_{r2}, w_{r2} are the values of the velocity rotational component in an individual fullerene carbon atom in the second position.

Since new coordinates x_3, y_3, z_3 have been found, it is not difficult to calculate new values for the projections of the moments of forces and the components of the inertia tensor from them. As a result, all coefficients in the linear system of algebraic equations will be determined:

$$\left.\begin{aligned} A_3\omega_{x3} + F_3\omega_{y3} + E_3\omega_{z3} &= K_x^n + \tfrac{\Delta t}{2} L_{x2}, \\ F_3\omega_{x3} + B_3\omega_{y3} + D_3\omega_{z3} &= K_y^n + \tfrac{\Delta t}{2} L_{y2}, \\ E_3\omega_{x3} + D_3\omega_{y3} + C_3\omega_{z3} &= K_z^n + \tfrac{\Delta t}{2} L_{z2}, \end{aligned}\right\} \quad (19)$$

From the last system, using Cramer's rule, we find the projections of the angular velocity of the molecule in the third position of the time step. At the end of the third step, we must not forget to calculate the projections of the velocities for the center of mass:

$$u_{c3} = u_c^n + \frac{\Delta t}{2} U_{c2},\ v_{c3} = v_c^n + \frac{\Delta t}{2} V_{c2},\ w_{c3} = w_c^n + \frac{\Delta t}{2} W_{c2}. \quad (20)$$

When obtaining coordinates and velocities in the fourth position, not $\frac{\Delta t}{2}$, but a full time step is used; therefore:

$$\begin{aligned} &x_4 = x^n + \Delta t(u_{c3} + u_{r3}),\ y_4 = y^n + \Delta t(v_{c3} + v_{r3}),\ z_4 = z^n + \Delta t(w_{c3} + w_{r3}). \\ &u_{c4} = u_c^n + \Delta t U_{c3},\ v_{c4} = v_c^n + \Delta t V_{c3},\ w_{c4} = w_c^n + \Delta t W_{c3}. \end{aligned} \quad (21)$$

Here

$$\begin{aligned} u_{r3} &= (z_3 - z_{c3})\omega_{y3} - (y_3 - y_{c3})\omega_{z3}, \\ v_{r3} &= (x_3 - x_{c3})\omega_{z3} - (z_3 - z_{c3})\omega_{x3}, \\ w_{r3} &= (y_3 - y_{c3})\omega_{x3} - (x_3 - x_{c3})\omega_{y3}. \end{aligned}$$

$$\left.\begin{array}{l} A_3\omega_{x4} + F_4\omega_{y4} + E_4\omega_{z4} = K_x^n + \Delta t L_{x3}, \\ F_3\omega_{x4} + B_4\omega_{y4} + D_4\omega_{z4} = K_y^n + \Delta t L_{y3}, \\ F_3\omega_{x1} + D_1\omega_{y4} + C_4\omega_{z4} = K_z^n + \Delta t L_{z3}, \end{array}\right\} \quad (22)$$

Then the coordinate values at the new time step can be found using the formulas:

$$\begin{array}{l} x^{n+1} = x^n + \frac{\Delta t}{6}[(u_{c1} + u_{r1}) + 2(u_{c2} + u_{r2}) + 2(u_{c3} + u_{r3}) + (u_{c4} + u_{r4})], \\ y^{n+1} = y^n + \frac{\Delta t}{6}[(v_{c1} + v_{r1}) + 2(v_{c2} + v_{r2}) + 2(v_{c3} + v_{r3}) + (v_{c4} + v_{r4})], \\ z^{n+1} = z^n + \frac{\Delta t}{6}[(w_{c1} + w_{r1}) + 2(w_{c2} + w_{r2}) + 2(w_{c3} + w_{r3}) + (w_{c4} + w_{r4})]. \end{array} \quad (23)$$

Here

$$\begin{array}{l} u_{r4} = (z_4 - z_{c4})\omega_{y4} - (y_4 - y_{c4})\omega_{z4}, \\ v_{r4} = (x_4 - x_{c4})\omega_{z4} - (z_4 - z_{c4})\omega_{x4}, \\ w_{r4} = (y_4 - y_{c4})\omega_{x4} - (x_4 - x_{c4})\omega_{y4}. \end{array}$$

The aim of the study is to research the character of fullerene C_{60} motion in the potential field of van der Waals forces of a stationary cylindrical nanocontainer. In turn, fullerene is a free object that has the ability to move inside the container and has kinetic energy. To check the accuracy of the calculation, we observed the balance of the system total mechanical energy:

$$H = \frac{J\omega^2}{2} + \frac{Mv^2}{2} + \sum_{i=1}^{N}\sum_{k=1}^{K} U(\rho_{ik}) = const, \quad (24)$$

$$J = A\alpha^2 + B\beta^2 + C\gamma^2 + 2D\beta\gamma + 2E\gamma\alpha + 2F\alpha\beta, \quad (25)$$

$$\alpha = \frac{\omega_x}{\omega},\ \beta = \frac{\omega_y}{\omega},\ \gamma = \frac{\omega_z}{\omega},\ \omega = \sqrt{\omega_x^2 + \omega_y^2 + \omega_z^2}. \quad (26)$$

Here, J is the moment of inertia relative to the instantaneous axis of rotation; ω is the instantaneous value of the angular velocity of the fullerene; M is the fullerene mass; v is the velocity of the center of mass in the fullerene; ρ_{ik} are the distances between atoms belonging to different molecules—the fullerene and nanocontainer.

4. Results and Discussion

The nature of the motion of C_{60} fullerene in a cylindrical stationary carbon nanocontainer with flat rectangular graphene covers was studied by calculation. We followed the trajectory of the center of mass of the fullerene and the projections of its angular velocities on the axes of the absolute basis. It should be noted that at small intervals of motion, the energy exchange between the container body and the fullerene is insignificant. Therefore, up to a certain point in time, the system can be considered Hamiltonian. However, some general trends can be identified at this initial stage of development in the process.

The balance of total mechanical energy is shown in Figure 2. It can be seen that the relative value of the error in calculating the Hamilton function is 5×10^{-4}% or 5×10^{-6}% relative units. This is a high accuracy in calculations, if we take into account that multiple fullerene impacts on the container walls are realized.

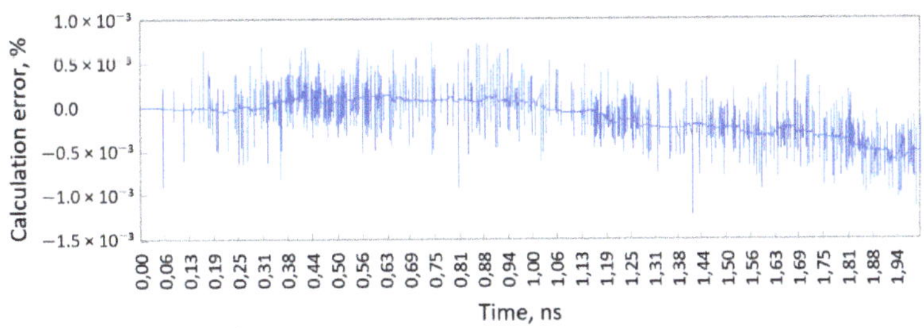

Figure 2. Calculation error in calculating the total mechanical energy in the system.

At the initial moment of time, the fullerene is in the center of the container, its center of mass is at rest and the angular velocity vector had a direction along the container axis: $\omega_x^0 = 0$, $\omega_y^0 = 0$, $\omega_z^0 = 100\ \text{ns}^{-1}$. Since this position is not an equilibrium position, the fullerene begins its movement towards the zones of most probable occurrence. Such zones are potential wells near the walls of the container.

In order to understand the nature of fullerene motion, the entire calculated time interval was divided into successive intervals. On the left in each of the figures below, the projection of the trajectory of the center of mass of the fullerene onto the *xy* plane is shown and on the right, onto the *xz* plane. Both of these flat figures in their totality give some idea of the spatial character of the fullerene motion. Figure 3 shows the very initial phase of the buckyball movement.

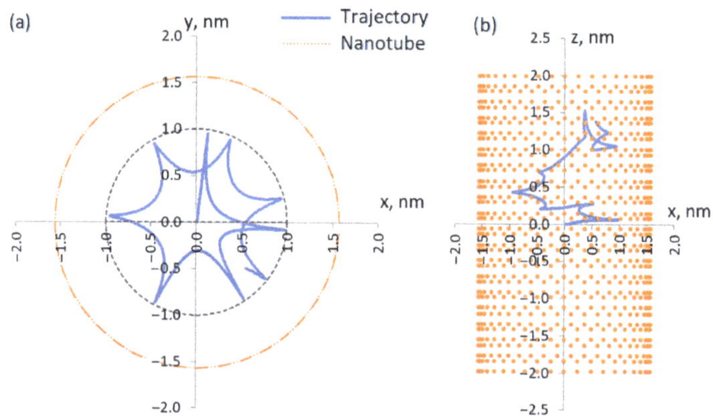

Figure 3. Trajectory of the fullerene center of mass (movement time up to 0.2 ns) in a nanotube:(**a**) in the *xy* plane, (**b**) in the *xz* plane.

It can be concluded from the figure that the fullerene initially falls almost radially onto the side surface of the container, i.e., moves to the nearest potential well. Over time, the action of the end potential well increases and the buckyball, along with radial and circumferential displacements, acquires speed in the axial direction. If we take into account the axial projection of the fullerene trajectory (left images in Figures 4–7), we clearly see fullerene jumps from one position to another. Since the presented mathematical model already includes an ideally elastic interaction with the container walls, the main element in the buckyball movement near the barrier and the carbon material is the jump. In other words, the fullerene in the nanocontainer bounces, like a tennis ball hitting a hard surface. At the same time, it is not in the field of gravitational influence, but in the field of van der Waals forces, which have a resulting direction towards the closest walls of the container.

Figure 4 shows motion over a time interval of up to 0.5 ns.

At this time interval, there is a decrease in the height of jumps and an increase in their frequency. Moreover, the movements of the buckyball are realized in the potential wells of the side surface of the nanocylinder.

In the next interval from 0.5 ns to 1.0 ns (Figure 5), the motion remained the same, but a zone of thickening for the trajectories appeared near the top cover of the nanocylinder.

At the time intervals shown in Figures 6 and 7, the nature of the movement changed somewhat: buckyball jumps began to occur at the end surfaces. Diametrical and almost diametrical crossings also appeared here, after which there is a transition along the side surface of the container on the other cover, where the fullerene enters a new end zone of attraction. In this regard, the most likely areas for the presence of a buckyball in the container are potential wells near flat covers.

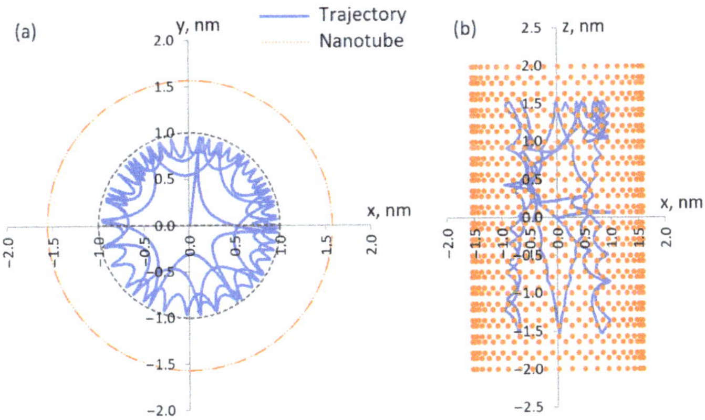

Figure 4. Trajectory of the fullerene center of mass (movement time up to 0.5 ns) in a nanotube: (**a**) in the *xy* plane, (**b**) in the *xz* plane.

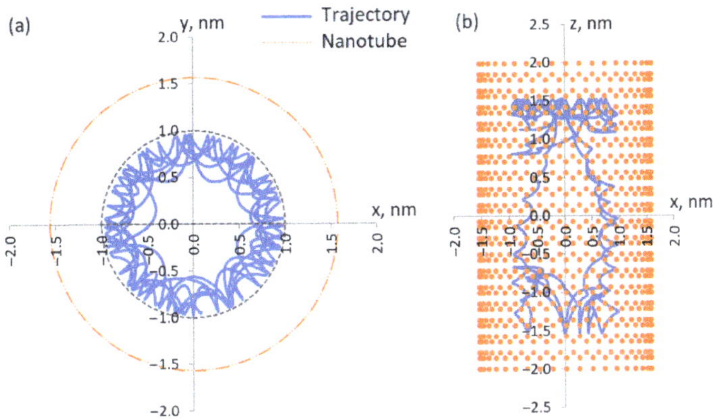

Figure 5. Trajectory of the fullerene center of mass (movement time from 0.5 to 1 ns) in a nanotube: (**a**) in the *xy* plane, (**b**) in the *xz* plane.

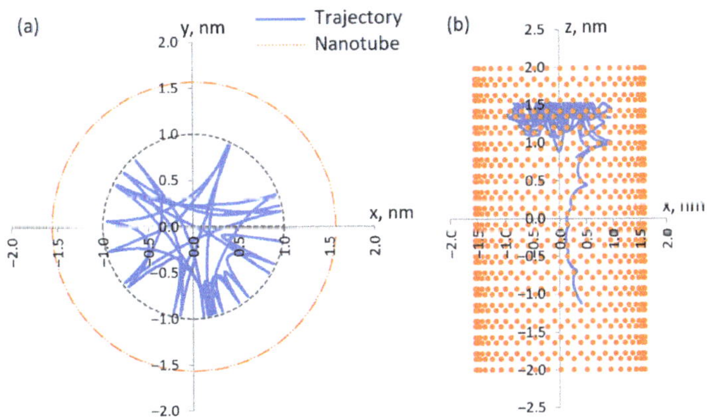

Figure 6. Trajectory of the fullerene center of mass (movement time from 1 to 1.5 ns) in a nanotube: (**a**) in the *xy* plane, (**b**) in the *xz* plane.

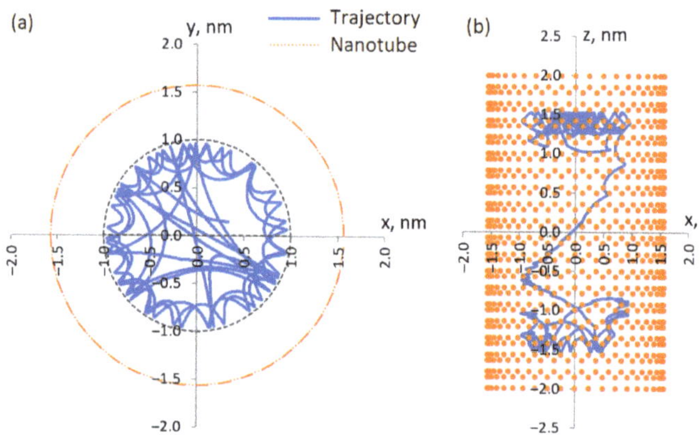

Figure 7. Trajectory of the fullerene center of mass (movement time from 1.5 to 2 ns) in a nanotube: (a) in the xy plane, (b) in the xz plane.

In the process of fullerene moving along the admissible positions inside the nanocontainer, a significant change in its kinetic energy components occurs. For example, in potential wells near the walls, the speed of the center of mass in the buckyball and its angular velocity noticeably increase. In this case, the fullerene rotation energy changes abruptly, immediately after approaching the limiting surface of the container and after approaching; as before, it has a constant value (Figure 8a).

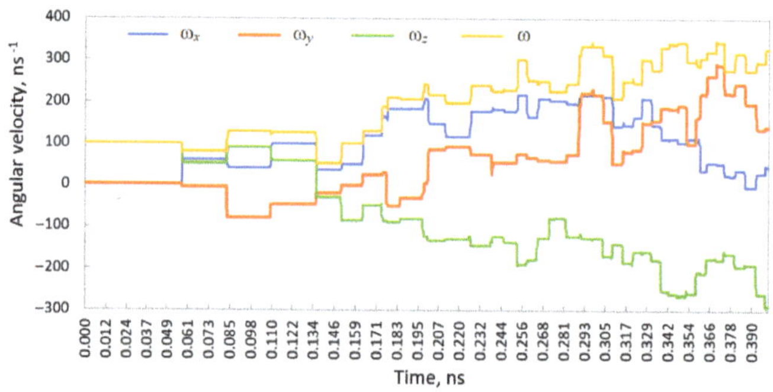

(a)

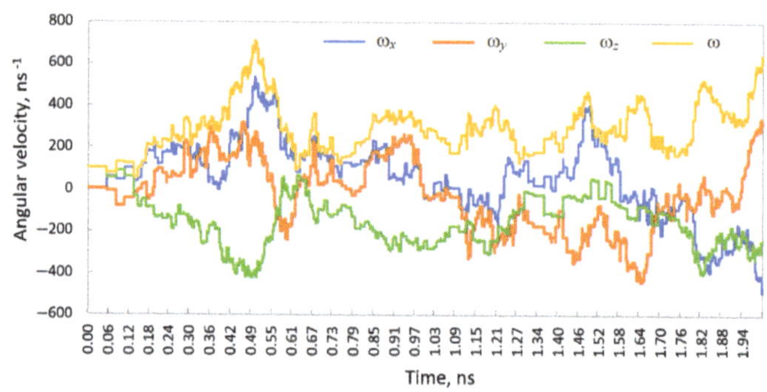

(b)

Figure 8. Projections of the angular velocity vector (initial value $\omega_x^0 = 0$, $\omega_y^0 = 0$, $\omega_z^0 = 100\ \mathrm{ns}^{-1}$) on the coordinate axis: (a) in case of movement time up to 0.4 ns; (b) in case of movement time up to 2 ns.

In other words, the rotational energy is quantized because the angular velocity is quantized, both in magnitude and in direction. This stepwise transition is realized with each hit against the wall. The same thing happens with an ordinary ball or a billiard ball when it interacts with the sides. The graph in Figure 8b shows the periodic nature of the global change in rotation speed.

5. Conclusions

The developed approach makes it possible to describe, in detail, any fullerene motions, in particular, its rotations and displacements with multiple impacts against the walls of the nanocontainer. As a result of these impacts, the energy of rotation changes abruptly. At the same time, it has constant values between beats. The main way to move the fullerene inside the container is to jump from one position on the container walls to another. When the fullerene center enters the potential well and moves predominantly in a direction tangential to the walls, the height of the jumps decreases and the frequency increases. Potential wells near the flat covers of the container are the most probable areas for the presence of fullerene, which is initially located in the center of the container.

The proposed approach for describing the motion of fullerenes is valid for any large and unchanging molecular structures during the motion. Moreover, it is valid for describing the rotations of the so-called "solid" bodies. For this, however, it is necessary to have tabulated values of the axial and centrifugal moments of inertia in the body, previously found for all possible directions of the coordinate axes.

Author Contributions: Conceptualization, methodology and project administration, A.M.B.; software, validation and investigation, D.V.M.; resources and data curation, A.V.L.-F.; formal analysis, writing—original draft preparation, writing—review and editing, A.M.B. and A.S.Ch.; visualization, A.S.Ch.; supervision and funding acquisition, M.A.B. All authors have read and agreed to the published version of the manuscript.

Funding: This research was funded by the Ministry of Science and Higher Education of Russia, grant number 075-02-2022-884.

Institutional Review Board Statement: Not applicable.

Informed Consent Statement: Not applicable.

Data Availability Statement: Not applicable.

Conflicts of Interest: The authors declare no conflict of interest. The funders had no role in the design of the study; in the collection, analyses, or interpretation of data; in the writing of the manuscript, or in the decision to publish the results.

References

1. Ma, F.; Li, Z.; Xu, H.; Li, Z.; Wu, D.; Li, Z.; Gu, F.L. Proton transfer in the complex H_3N center dot center dot center dot HCl catalyzed by encapsulation into a C_{60} cage. *ChemPhysChem.* **2009**, *10*, 1112–1116. [CrossRef] [PubMed]
2. Kurotobi, K.; Murata, Y. A single molecule of water encapsulated in fullerene C_{60}. *Science* **2011**, *333*, 613–616. [PubMed]
3. Komatsu, K.; Murata, M.; Murata, Y. Encapsulation of molecular hydrogen in fullerene C_{60} by organic synthesis. *Science* **2005**, *307*, 238–240. [PubMed]
4. Farimani, A.B.; Wu, Y.B.; Aluru, N.R. Rotational motion of a single water molecule in a buckyball. *Phys. Chem. Chem. Phys.* **2013**, *15*, 17993–18000. [PubMed]
5. Joseph, S.; Aluru, N.R. Why are carbon nanotubes fast transporters of water? *Nano Lett.* **2008**, *8*, 452–458. [PubMed]
6. Heiranian, M.; Farimani, A.B.; Aluru, N.R. Water desalination with a single-layer MoS_2 nanopore. *Nat. Commun.* **2015**, *6*, 8616. [PubMed]
7. Suk, M.E.; Aluru, N.R. Water Transport through ultrathin graphene. *J. Phys. Chem. Lett.* **2010**, *1*, 1590–1594.
8. Farimani, A.B.; Heiranian, M.; Aluru, N. Nano-electro-mechanical pump: Giant pumping of water in carbon nanotubes. *Sci. Rep.* **2016**, *6*, 26211. [CrossRef] [PubMed]
9. Alipour, A.; Ansari, R.; Sadeghi, F. Oscillation of C_{60} fullerene in carbon nanotube bundles. *J. Vib. Acoust.* **2013**, *135*, 051009.
10. Kang, J.W.; Lee, K.W. Molecular dynamics study on the C_{60} oscillator in a graphene nanoribbon trench. *J. Korean Phys. Soc.* **2014**, *65*, 185–189. [CrossRef]

11. Ulbricht, H.; Moos, G.; Hertel, T. Interaction of C_{60} with carbon nanotubes and graphite. *Phys. Rev. Lett.* **2003**, *90*, 095501. [PubMed]
12. Wu, C.; Fang, T.; Chan, C. A molecular dynamics simulation of the mechanical characteristics of a C_{60}-filled carbon nanotube under nanoindentation using various carbon nanotube tips. *Carbon* **2011**, *49*, 2053–2061.
13. Zhou, L.; Zhu, B.; Pan, Z.Y.; Wang, Y.X. Reduction of the buckling strength of carbon nanotubes resulting from encapsulation of C_{60} fullerenes. *Nanotechnology* **2007**, *18*, 275709.
14. Tu, G.; Guo, D.; Li, B.; Zhang, H. Transport properties and mechanism of C_{60} coupled to carbon nanotube electrode. *Phys. B Condens. Matter* **2011**, *406*, 2138–2142. [CrossRef]
15. Shokri, A.; Nikzad, S. Electrical transport properties of a CNT/C_{60}/CNT hybrid junction with closed end CNT leads using Green's function method. *Eur. Phys. J. B* **2013**, *86*, 22. [CrossRef]
16. Ning, G.; Kishi, N.; Okimoto, H.; Shiraishi, M.; Kato, Y.; Kitaura, R.; Sugai, T.; Aoyagi, S.; Nishibori, E.; Sakata, M.; et al. Synthesis, enhanced stability and structural imaging of C_{60} and C_{70} double-wall carbon nanotube peapods. *Chem. Phys. Lett.* **2007**, *441*, 94–99. [CrossRef]
17. Hirahara, K.; Suenaga, K.; Bandow, S.; Kato, H.; Okazaki, T.; Shinohara, H.; Iijima, S. One-dimensional metallofullerene crystal generated inside single-walled carbon nanotubes. *Phys. Rev. Lett.* **2000**, *85*, 5384. [CrossRef] [PubMed]
18. Ghavanloo, E.; Fazelzadeh, S.A.; Rafii-Tabar, H. A computational modeling of Raman radial breathing-like mode frequencies of fullerene encapsulated inside single-walled carbon nanotubes. *J. Mol. Model.* **2017**, *23*, 48. [PubMed]

crystals

MDPI

Article

The Exchange-Correlation Effects on the Electronic Bands of Hybrid Armchair Single-Walled Carbon Boron Nitride Nanostructure

Yahaya Saadu Itas [1], Abdussalam Balarabe Suleiman [2], Chifu E. Ndikilar [2], Abdullahi Lawal [3], Razif Razali [4], Mayeen Uddin Khandaker [5,*], Pervaiz Ahmad [6], Nissren Tamam [7] and Abdelmoneim Sulieman [8]

1 Department of Physics, Bauchi State University Gadau, PMB 65, Gadau 751105, Nigeria; yitas@basug.edu.ng
2 Department of Physics, Federal University Dutse, Dutse 720101, Nigeria; salam@fud.edu.ng (A.B.S.); chifu.ndikilar@fud.edu.ng (C.E.N.)
3 Department of Physics, Federal College of Education, Zaria 810282, Nigeria; labdullahi2@live.utm.my
4 Department of Physics, Faculty of Science, Universiti Teknologi Malaysia, Skudai 81310, Malaysia; razifrazali@utm.my
5 Centre for Applied Physics and Radiation Technologies, School of Engineering and Technology, Sunway University, Bandar Sunway 47500, Malaysia
6 Department of Physics, University of Azad Jammu and Kashmir, Muzaffarabad 13100, Pakistan; pervaiz.ahmad@ajku.edu.pk
7 Department of Physics, College of Sciences, Princess Nourah bint Abdulrahman University, P.O. Box 84428, Riyadh 11671, Saudi Arabia; nmtamam@pnu.edu.sa
8 Department of Radiology and Medical Imaging, College of Applied Medical Sciences, Prince Sattam Bin Abdulaziz University, P.O. Box 422, Alkharj 11942, Saudi Arabia; a.sulieman@psau.edu.sa
* Correspondence: mayeenk@sunway.edu.my

Citation: Itas, Y.S.; Suleiman, A.B.; Ndikilar, C.E.; Lawal, A.; Razali, R.; Khandaker, M.U.; Ahmad, P.; Tamam, N.; Sulieman, A. The Exchange-Correlation Effects on the Electronic Bands of Hybrid Armchair Single-Walled Carbon Boron Nitride Nanostructure. *Crystals* **2022**, *12*, 394. https://doi.org/10.3390/cryst12030394

Academic Editors: Walid M. Daoush, Fawad Inam, Mostafa Ghasemi Baboli and Maha M. Khayyat

Received: 13 February 2022
Accepted: 12 March 2022
Published: 14 March 2022

Publisher's Note: MDPI stays neutral with regard to jurisdictional claims in published maps and institutional affiliations.

Abstract: This study investigates the effect of exchange-correlation on the electronic properties of hybridized hetero-structured nanomaterials, called single-walled carbon boron nitride nanotubes (SWCBNNT). A first principles (ab initio) method implemented in Quantum ESPRESSO codes, together with different parametrizations (local density approximation (LDA) formulated by Perdew Zunga (PZ) and the generalized gradient approximation (GGA) proposed by Perdew–Burke–Ernzerhof (PBE) and Perdew–Wang 91 (PW91)), were used in this study. It has been observed that the disappearance of interface states in the band gap was due to the discontinuity of the π–π bonds in some segments of SWCNT, which resulted in the asymmetric distribution in the two segments. This work has successfully created a band gap in SWCBNNT, where the PBE exchange-correlation functional provides a well-agreed band gap value of 1.8713 eV. Effects of orbitals on electronic properties have also been studied elaborately. It has been identified that the P^y orbital gives the largest contribution to the electrical properties of our new hybrid SWCBNNT nanostructures. This study may open a new avenue for tailoring bandgap in the hybrid heterostructured nanomaterials towards practical applications with next-generation optoelectronic devices, especially in LED nanoscience and nanotechnology.

Keywords: SWCBNNT heterostructures; hybrid system; quantum ESPRESSO; band gap; GGA functionals

1. Introduction

Carbon nanotubes (CNTs) have been evolved as important materials for the advancement of nanoscience and technology [1]. They have received much attention because of their ability to behave as both metallic and semi-metals depending on the chirality/translation of carbon atoms arranged in a hexagonal lattice [2]; they can be single-walled (SW) or multi-walled (MW) structures. They are anisotropic and also exist in three different geometries as an armchair, zigzag, and chiral. To bring the CNTs to the next-generation optoelectronic fields or applications, many research approaches comprising theoretical [3], computational, and experimental [4] ideas are carried out, and they all come up with many exciting results [5,6]. A wide range of potential applications of CNTs in the field of polymer, composites, hydrogen/energy storage, biomedical sciences, field emission dipoles, etc., were also reported. Although CNTs show unique structural and physical properties, they still require further improvement to be used in certain fields of nanotechnology [7]. As a result, CNTs hetero-structures have become one of the most interesting areas where scientists are trying to explore novel properties for application purposes. The CNT composites such as carbon nanotube metal matrix composites (CNT-MMC) are formed to make alloys due to the CNT's high tensile strength and electrical conductivity. Copper/carbon nanocomposites are produced to fill the high demand for copper substitutes [8]. CNT composites are used as reinforcement and thermal reservoirs. The high demand for semiconductor devices today has led us to develop the idea of combining SWCNT with a wide gap material such as boron nitride nanotube (BNNT) to form a CNT–BNNT hetero-structure. The hybrid single-walled carbon boron nitride nanotube (SWBNNT) has an identical hexagonal structure to that of a single-walled carbon nitride nanotube (SWCNT). It is worth mentioning that a single-walled boron nitride nanotube (SWBNNT) is an insulator or wide bandgap semiconductor with an energy gap of 5–6 eV [9], whereas armchair SWCNT is a conductor in its pure form [10]. Therefore, the idea of creating a carbon boron nitride nanotube heterostructure may create an energy band gap in CNT which will reduce its electrical conductivity to the level of semiconductors. Just like CNTs, the BNNT is thermally and chemically inert, and is known for its anisotropic behavior [11]. It is an isomorph of graphene [12]. Generally, armchair SWCNTs are rarely used as pure semiconductors because of compatibility problems that arose from their hexagonal structure and isotropic nature [13]. SWCNTs are used as semiconductors when they are fabricated in an impure form, where the intrinsic impurities result in the creation of a band gap that reduces and/or tries to terminate the full electrical behavior of the nanotubes. Successful creation of CBNNT hybrid nanotubes can only be achieved when we use a nanotube of the same structural properties, such as crystallographic nature (hexagonal structure), anisotropy, piezoelectricity, pyroelectricity, and biocompatibility. A successful attempt was made to create a band gap in CNT with the optimized structure of bilayered tin selenide (SnSe) [14] by using the Quantum ESPRESSO package. Tin selenide (SnSe) has a hexagonal honeycomb structure similar to graphene with a separation of 1.56 Å between Sn and Se atoms [14–16]. The result was the creation of a SnSe-CNT semiconductor with a narrow bandgap of 2.56 eV. The Perdew–Burke–Ernzerhof (PBE) type of generalized gradient approximation (GGA) exchange-correlation was used to obtain the said result. Following the same strategy, as demonstrated in the literature [17], we have been motivated to study the electronic band structures for hybrid SWCBNNT structures. Consequently, in this work, we have implemented the ab initio principles to calculate the electronic properties of our newly created system of carbon boron nitride nanotubes (CBNNT) and analyze its potential in the next generation semiconductor applications. Some of the relevant works on the effect of exchange correlations and band gaps obtained both theoretically and experimentally are summarized in Table 1.

Table 1. Available studies on band gap in hybrid nanostructures by various methods.

DFT Method	Theoretical Results	Experimental Results	References
DFT-theoretical	Reported the possibility of obtaining band gap by combining CNT segments and BNNT segments.	None	[18]
DFT-Theoretical	It has been reported that the CNT-BNNT can be competitive in thermodynamical stability for sufficiently large segments of building blocks in the axial direction.		[19]
Non-equilibrium Green's function method combined with the density functional theory	Carbon and boron nitride nanotubes were obtained with semiconducting properties of 2.2 eV; results predicted that CBNNT could become potential candidates in the field of nano rectifiers.		[20]

Table 1. *Cont.*

DFT Method	Theoretical Results	Experimental Results	References
Density functional theory and using basis set 6–31 g (d,p)	Reported the band gap of 1.21 and 2.52 eV which is close with our results of PW91.		[21]
Geometry optimization implemented in the CASTEP package	Reported 2.3 eV band gap in hetero nanotubes with the lowest unoccupied molecular orbital and the highest occupied molecular orbital mainly located on the carbon nanotube section.		[22]
Vienna ab initio simulation package	The electrical conductivity of CBNNT is increased by oxygen absorption.		[23]
CVD		2.41 eV band gap was reported. Recommended for theoretical and computational confirmatory tests.	[24]
VASP code	Reported 1.06 eV band gap, also reported that the highest occupied and lowest unoccupied orbital gap of carbon-boron-nitride hetero nanotubes can be significantly tuned by modifying the CNT and BNNT general geometry.		[25]
LDA	Reported 5.6 eV band gap in BNNT. Furthermore, analysis of the HOMO–LUMO gap after the adsorption process showed that the HOMO value increased marginally while the LUMO value decreased dramatically in the curcumin-BNNT complexes		[26]
GGA-PBE	Reported 1.83 eV band gap, which agrees with this current research. Also, reported that the band gap of the CBNNT system is greatly influenced by the nanotube aspect ratio.		[27]
GGA-PW91	Reported 2.52 eV band gap, highlighted the potentials of CBNNT for the next generation spintronics.		[28]

As can be seen in Table 1, various methods have been used to study CBNNT nanostructures, and various results have been obtained. In our work, a novel method of inter-tube coupling (which is the basis of this research) was used (Figure 1). As far as our concern, our technique was not adopted in any of the previous studies available in the literature. As such, this work forms a new pathway for tailoring band gap in hybrid heterostructured nanomaterials for advanced optoelectronic applications.

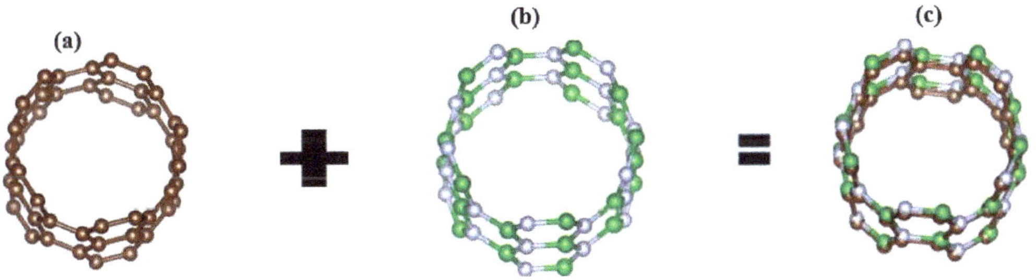

Figure 1. Hybrid SWCBNNT hetero-nanotubes optimized by inter-tube coupling. (**a**) Metallic (5, 5) SWCNT (**b**) Non-metallic (5, 5) SWBNNT and (**c**) Semi-conducting (5, 5) SWCBNNT.

2. Computational Methods

This section demonstrates the adopted working methodology that has made possible the successful implementation of this study. The method shows effectiveness in developing a new set of processes that could eventually help the relevant researchers for creating a new band gap in the potential hetero-combination of BNNT-CNT nanostructures. In this study, a representative model of both (5, 5) armchair SWBNNT and SWCNT was considered. The local density of state (LDOS), the total density of state (DOS), and the electronic band's structure of the armchair form of CBNNT hetero-nanotubes are calculated within the self-consistent field through solving the Kohn–Sham equation within the DFT in terms of LDA and GGA functional (a method implemented on Quantum ESPRESSO codes), this is necessary in order to analyze the effect of the various exchange-correlation functional on the bands' alignment of our hetero-system. The Quantum ESPRESSO (an acronym for open-Source Package for Research in Electronic Structure, Simulation, and Optimization) is an integrated suite of open-source computer codes for electronic-structure calculations and materials modeling at the nanoscale. It is based on density-functional theory, plane waves, and pseudopotentials. It uses first-principles electronic-structure calculations and materials modeling, distributed for free and as free software under the GNU General Public License.

Calculations are performed on the $2 \times 2 \times 1$ supercell model of (5, 5) SWBNNT primitive unit cell containing 80 carbon atoms and 16 atoms each of boron and nitrogen, based on the first principles together with DFT implemented in quantum ESPRESSO codes. The calculations for the exchange-correlation are performed within the Perdew-Zunga (PZ) within LDA, Perdew–Burke–Ernzerhof (PBE), and Perdew–Wang 91 (PW91) of GGA approximations together with the smearing occupations method of integrals. We have determined the Brillouin zone by using the Monkhorst–Pack scheme with k-grids of $1 \times 1 \times 4$ and an e-cut of 50 Ry.

3. Geometry Optimization

Because of the structural influence of the hybrid SWCBNNT system on the electronic properties, optimization of the tube geometry was performed prior to the calculation of the electronic properties. This is achieved with the codes implemented in the nanotubes modeler and VESTA. The hetero-nanotube was optimized by a new method called inter-tube coupling (Figure 1) in which the carbon atoms in CNT are coupled with boron and nitrogen atoms in BNNT.

The studies were conducted on the SWCBNNT system and the inter-tube separations were chosen as 3.95 Å, 4.80 Å, and 5.29 Å respectively. To ensure accurate results in this research, the nanotube was appropriately relaxed to appropriate geometries. In the SWCBNNT, the tube length and the tube height were chosen as 6.23 Å and 4.26 Å, respectively. The chiral/translation vectors were constructed such that $n = 5$, $m = 5$ to ensure the proper armchair chirality. The results of the relax calculations are listed in Table 2. The maximum force, stress, and displacements were set at 0.06 eV/Å, 0.06 GPa, and 6×10^{-4} Å, respectively. The unit cell volume was 6515.67 Å^3 with lattice parameters a = 18.68 Å and c = 9.68 Å. As can be seen from iteration 3, there was zero error in the relax calculations; as such, the data in iteration 3 were used to obtain a well-converged value of all parameters used in this research.

Table 2. The geometry of SWCBNNT (5, 5).

Iterations	Delta-h	Delta-r	K-Point (Gamma)			Iterations		
						a_1	a_2	a_3
0			1.000000	1.000000	1.000000	41	41	0
1	1.7171×10^{-3}	7.5733×10^{-2}	1.001208	1.001208	1.000000	20	20	0
2	-4.1446×10^{-5}	-1.8236×10^{-4}	1.001205	1.001205	1.000000	0	0	0
3	0.0000e + 00	0.0000e + 00	1.001205	1.001205	1.000000			

4. Results and Discussion

4.1. Bands Structures of (5, 5) CBNNT under Three Different Exchange-Correlation Functionals

The electronic band structures of the system of (5, 5) armchair single-walled carbon boron nitride nanotubes (SWCBNNT) hybrid material were studied, and the results are compared under three different exchange-correlation functional implemented within the local density approximation (LDA) and the generalized gradient approximation (GGA). The results obtained with each pseudopotential showed that direct band gaps [29] were obtained, with LDA-PZ pseudo potential being the lowest of 0.0433 eV at the gamma point. A considerable improvement of the band gap was obtained when PBE pseudopotential was used for our calculations. As can be seen in Figures 2b and 3b, a direct band gap of 1.87 eV was achieved with PBE and 0.1886 with PW91 (as presented in Figures 2c and 3c. The result obtained with PBE is close to the value of 2.00 eV experimental value [30]. This value of band gap makes the CBNNT behave as a semi-metal with a tunable direct band gap [31], a property that can be applied in LED, spintronic, electronic, Schottky devices, and photonics devices with tunable band structures [32].

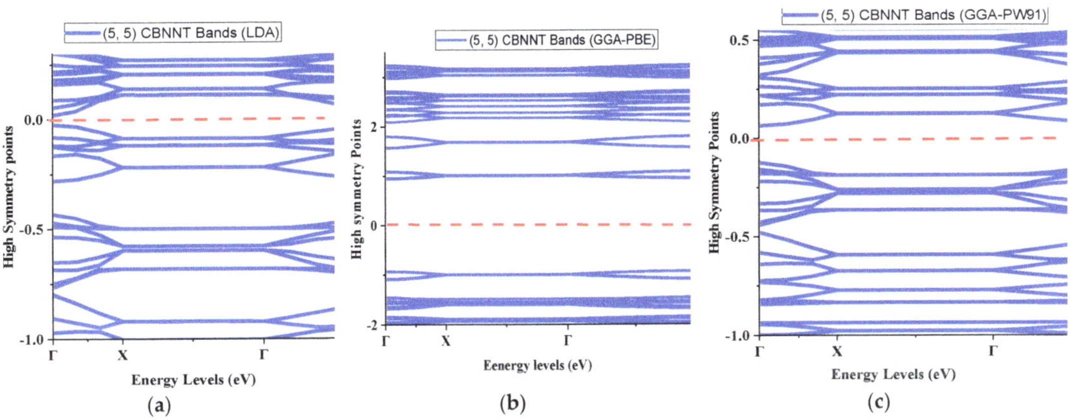

Figure 2. Bands structures of (5, 5) SWCBNNT (**a**) LDA (**b**) PBE (**c**) PW91.

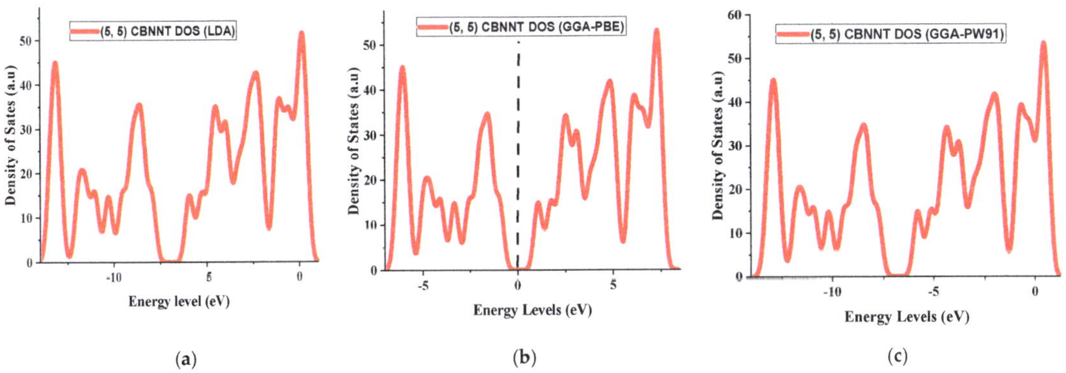

Figure 3. TDOS for (5, 5) CBNNT systems under three different exchange-correlations of LDA and GGA. (**a**) Effect of LDA; (**b**) effect of PBE; (**c**) effect of PW91.

The effect of each pseudopotential is summarized in Table 3. It can be seen that different total energies were achieved by various exchange-correlation functional; for example, the total energy was −12.99 Ry with PZ functional. This gives a Fermi energy of −8.16 eV in 72 self-consistent fields (SCF) iterations. However, different total energy and Fermi energy were obtained by PBE functional, although the SCF calculation was achieved in the same 72 iterations. In the case of PW91, the total energy achieved was −1314.24 Ry (i.e., Rydberg unit of energy) and the Fermi energy was −8.00 eV. Since the highest band gap was achieved with PBE exchange functional, we consider it as the optimum parameter of the generalized gradient approximation in solving the exchange-correlation problem, because it is reported in previous studies that the PBE is faster than the other exchange-correlation [33].

Table 3. The effect of exchange-correlation in the Fermi energy.

S/No	Pseudopotential	Total Energy Achieved (Ry)	The Calculated Band Gap (eV)	SCF Iterations
1	LDA	−1299.17	0.043	72
2	PBE	−1444.79	1.87	72
3	PW91	−1314.23	0.19	74

4.2. *Analysis of the Density of States*

To verify more on the findings regarding the bands obtained, analyses were made on the density of states and partial density of states for our SWCBNNT hybrid system. As can be seen in Figure 3a,b, there are more states in the valence band than in the conduction band for bands obtained with PZ and PW91 exchange functional. This is because the total energy of the nanotube that contributes to the conduction is lower than the sum of the energy which forms the hetero-nanotube. As a result, the total energy of the nanotube which contributes to conduction is lower than the sum of the energy which forms the hetero-nanotube. Direct bands are obtained because the highest energy of the valence band is equal to the lowest energy of the conduction band. It means that they are at the same momentum when the transition takes place [34]. Recombination of holes and electron takes place in order to conserve the momentum energy that is released in the form of light, such as LED. A good band can be seen in Figure 3b when PBE exchange functional was used. Although the PBE underestimates band gaps, this value is close to the 2.0 eV experimental value reported elsewhere [35]. Figure 3b gives the plot information of TDOS for (5, 5) SWCBNNT system within the PBE implementation. Zero states can be seen at the Fermi level; as such, this region can be considered as the gap calculated to be 1.8713 eV. Very few states can be seen in Figure 3a,b. There are also more states in the conduction band than in the valence band. The three highest peaks can be seen with different states. The first one, at −6.109 eV, is due to domination by $1S^2$ orbital of boron and carbon atoms, respectively (refer to Figures 4a,b and 5a,b). The second peak, at −1.709 eV, is due to collective contributions by $2S^2$ orbitals of B, C, and N atoms, respectively [36]. Lower occupations are due to partial contributions by S orbitals of all the systems (refer to Figure 4). The third state occurs at 7.291 eV. This is due to collective dominations by $2P^y$ orbitals of B, C, and N atoms. The presence of zero states at zero energy level confirms that the band gap had been successfully created by using PBE exchange functional. In the case of the PW91 exchange functional, dense states are seen in the valence bands because the total energy of interactions is the sum of exchange energy and interaction energy of the nanotube which contributes to the conduction band, and that is lower than the sum of the formation energy which forms the hetero-nanotube.

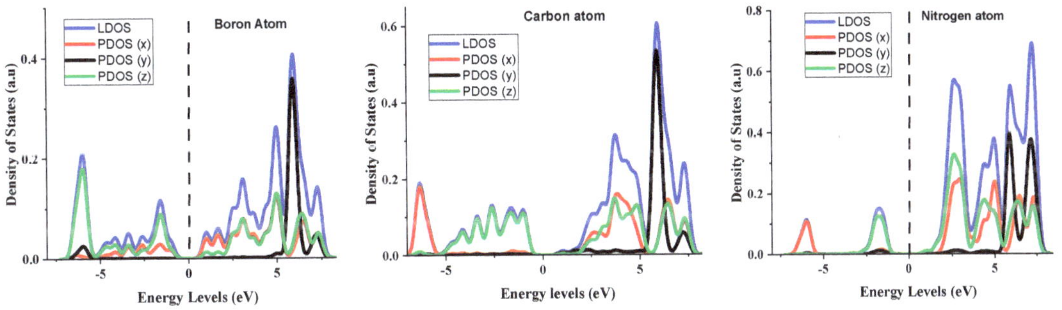

(a) (b) (c)

Figure 4. PDOS of (5, 5) CBNNT system under GGA-PBE. (**a**) Boron atom (**b**) carbon atom (**c**) Nitrogen atom.

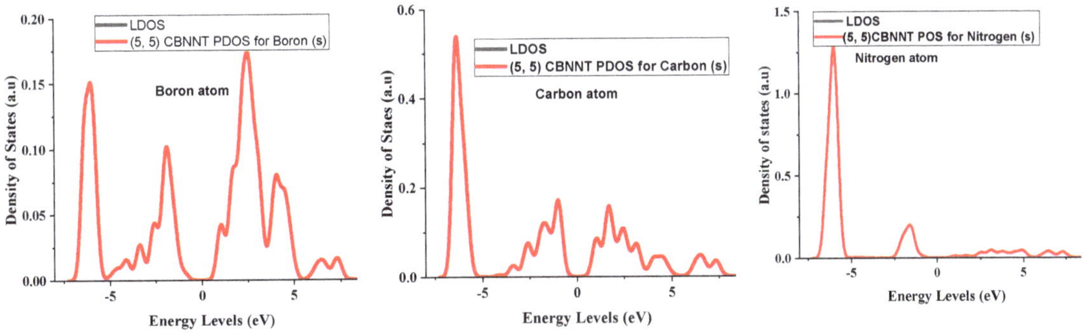

(a) (b) (c)

Figure 5. S orbital contribution to electrical properties of (5, 5) CBNNT system. (**a**) Boron atom (**b**) carbon atom (**c**) Nitrogen atom.

4.3. Partial Density of States for the (5, 5) CBNNT System

These studies of PDOS are limited to the results implemented with LDA and PBE exchange functional. The individual orbital's contributions and effect on the hybridized SWCBNNT system were studied. The SWCBNNT semiconductor created in this work is only a hybrid of S and P orbitals. Moreover, all the constituent elements belonging to the same group IV in the periodic table with only carbon, among them, can form a covalent bond to itself with electron mobility of 15,000 cm^2/Vs in the graphene lattice [37]. The studies of PDOS therefore may help to understand the electrical mobility of our semiconductor system. Figure 6 shows that P^y orbital generally determines the band gap of our SWCBNNT hybrid system because it gives the largest contribution in all the constituent elements. However, our investigation revealed that P^x orbital provided the smallest contribution to the conduction process. This is because they are considered frozen and form some part of the nucleus. They are also always filled with both up and down spins, hence requiring more energy of excitation [38]. The P^z orbitals are partially occupied with less probability of finding the electron. For example, the P^z orbital of carbon is having zero electrons; hence, will have to make a little contribution by interaction with one of the 2S electrons. The interaction between each P^z orbitals among the carbon atoms is due to the result of the π-bonding. Therefore, the electrical mobility across the Fermi energy is mainly by the carbon atoms in the carbon nanotube lattice. This can be justified from Figure 3, which shows that the P^z orbital of carbon contributes more than the P^z orbitals of boron and nitrogen atoms.

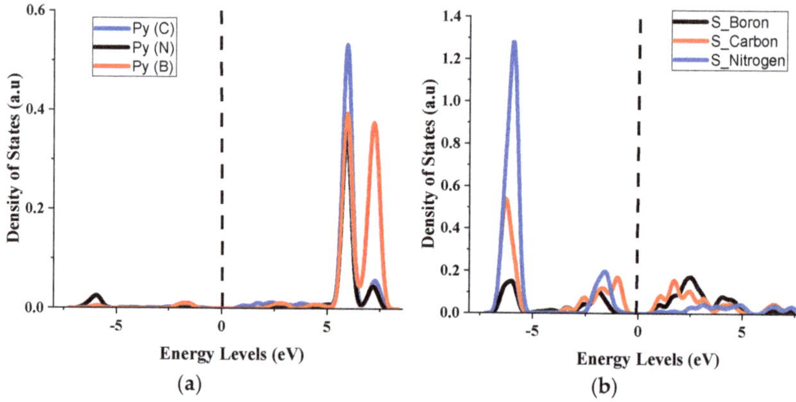

Figure 6. (**a**) P^y contributions for B, C, and N, respectively; (**b**) S contributions for B, C, and N, respectively.

There is a little contribution by the S orbital of carbon because of its interaction with carbon P^z orbitals. The S orbitals of other elements can be seen to make almost no contributions. To justify this claim, the PDOS and LDOS of Figure 5 can be seen to overlap, which means that the local density of states is the contribution of the S orbital.

4.4. Effects of P^y and S Orbitals in the Semiconductivity of (5, 5) CBNNT

To report which of the P^y orbitals takes more of the electronic properties of our system, we analyzed the results shown in Figure 6. This figure reveals that the P^y orbital of carbon in SWCNT provides more dominations than the P^y orbitals of boron and nitrogen in SWBNNT. Therefore, it can be inferred that P^y orbitals of carbon contribute to narrowing the wider gap made by boron nitride nanotubes, which pave the way for semiconductivity. The results of this work agreed closely with previous findings from different methods. For example, ref. [18] reported that it is possible to obtain a new set of semiconducting hetero nanotubes by interfacing layers of CNT and BNNT. This has been possible in this work. The 1.8 eV band gap reported by [27] agrees with the result of 1.8713 eV obtained with the same PBE in this work. Moreover, the reported value of 1.21 by [21] is approximately close to the obtained 0.8 eV in this work with the same PW91.

Figure 6a illustrates the individual p^y contributions by each of the constituent elements, although they all belong to the same group in the periodic table [39]. They differ in the contribution to the conduction process by each of their p-orbitals, which arises due to different occupations by p^z orbital [40]. For example, the p^z orbital for element boron contributes to the electronic configuration $1S^2\ 2S^2\ 2P_x{}^1\ 2P_y{}^0\ 2P_z{}^0$. As can be seen, the p^y orbital of boron contributes less because it has zero electrons. In the case of carbon with an electronic configuration of $1S^2\ 2S^2\ 2P_x{}^1\ 2P_y{}^1\ 2P_z{}^0$, it can be seen that the p^y orbital of carbon is occupied by one electron, hence making a higher contribution than boron. This can be seen in Figure 6b. All the S′ orbitals (Figure 6b) are found to make a very negligible contribution to the conduction process because they are considered frozen and form some part of the nucleus [41–46]. They are also always filled with both up and down spins, hence requiring more energy of excitation similar to p^x orbital.

5. Conclusions

In this study, the electronic structure of the hybridized armchair form of SWCBNNT was calculated based on the first principles method. This study successfully created a band gap under three different parameterizations of LDA and GGA using the density functional theory which is implemented in Quantum ESPRESSO code. In each case, the armchair configurations of carbon nanotubes and boron nitride nanotubes were chosen to form coupled carbon–boron–nitride nanotubes via a new method of inter-tube coupling. The chirality of all systems was chosen to be that of (5, 5) armchair tube. It was found that the band structure of our (5, 5) SWCBNNT is straightforward, and can be applied in the next-generation optoelectronic devices such as LED. The band gap of the hybrid SWCBBNNT system is seen to greatly depend on the orbital contributions to atomic interactions in which the p^y orbitals of carbon in SWCNT accounted for the electronic properties of the CBNNT hetero nanotubes material. This is because other p^x and p^z orbitals are considered to form some part of the nucleus which provide effective repulsion; hence, valence states are orthogonal to the core nuclear states and are considered frozen. Although the band gap of 1.8 eV has been recorded with PBE, it is well known that both LDA and GGA underestimate the band gap. This result, therefore, can be improved by using hybrid functions such as the GW functionals. The calculation of these band structures may hopefully lead to the addition of new knowledge in the literature, and then serve as a reference for further research on CBNNT heterostructures.

Author Contributions: Conceptualization, Y.S.I. and A.B.S.; methodology, Y.S.I. and C.E.N.; software, Y.S.I., C.E.N., R.R. and A.L.; formal analysis, Y.S.I. and A.B.S.; resources, A.S. and N.T.; data curation, P.A. and Y.S.I.; writing—original draft preparation, Y.S.I.; writing—review and editing, M.U.K.; visualization, P.A. and R.R.; funding acquisition, A.S. and N.T. All authors have read and agreed to the published version of the manuscript.

Funding: This work was supported by the Princess Nourah bint Abdulrahman University Researchers Supporting Project (Grant No. PNURSP2022R12), Princess Nourah bint Abdulrahman University, Riyadh, Saudi Arabia.

Institutional Review Board Statement: Not applicable.

Informed Consent Statement: Not applicable.

Data Availability Statement: No data reported.

Acknowledgments: The authors express their gratitude to Princess Nourah bint Abdulrahman University Researchers Supporting Project (Grant No. PNURSP2022R12), Princess Nourah bint Abdulrahman University, Riyadh, Saudi Arabia.

Conflicts of Interest: There is no conflict of interest to declare.

References

1. Wilson, M.; Evans, L. Carbon Nanotubes as Advanced Materials. *J. Aust. Ceram. Soc.* **2021**, *36*, 21–36.
2. Powell, L.R.; Kim, M.; Wang, Y. Chirality-Selective Functionalization of Semiconducting Carbon Nanotubes with a Reactivity-Switchable Molecule. *J. Am. Chem. Soc.* **2017**, *139*, 12533–12540. [CrossRef] [PubMed]
3. Melaibari, A.; Daikh, A.A.; Basha, M.; Abdalla, A.W.; Othman, R.; Almitani, K.H.; Hamed, M.A.; Abdelrahman, A.; Eltaher, M.A. Free Vibration of FG-CNTRCs Nano-Plates/Shells with Temperature-Dependent Properties. *Material* **2022**, *10*, 583. [CrossRef]
4. Janas, D. Special Issue of Materials Focused on "Electrical, Thermal and Optical Properties of Nanocarbon Materials". *Materials* **2022**, *15*, 1649. [CrossRef] [PubMed]
5. Foygel, M.; Morris, R.D.; Anez, D.; French, S.; Sobolev, V.L. Theoretical and computational studies of carbon nanotube composites and suspensions: Electrical and thermal conductivity. *Phys. Rev. B* **2015**, *71*, 1–6. [CrossRef]
6. Zhigilei, L.V.; Salaway, R.N.; Wittmaack, B.K.; Volkov, A.N. Computational Studies of Thermal Transport Properties of Carbon Nanotube Materials. In *Carbon Nanotubes for Interconnects*; Springer: Cham, Switzerland, 2017; pp. 129–161. [CrossRef]
7. Trivedi, M. Recent Development and Applications of Carbon Nanotubes. *Chem. Sci. Rev. Lett.* **2020**, *9*, 502–510. [CrossRef]
8. Sundaram, R.M.; Sekiguchi, A.; Sekiya, M.; Yamada, T.; Hata, K. Copper/carbon nanotube composites: Research trends and outlook. *R. Soc. Open Sci.* **2018**, *5*, 180814. [CrossRef]
9. Kim, K.S.; Kim, M.J.; Park, C.; Fay, C.C.; Chu, S.-H.; Kingston, C.T.; Simard, B. Scalable manufacturing of boron nitride nanotubes and their assemblies: A review. *Semicond. Sci. Technol.* **2016**, *32*, 3–13. [CrossRef]
10. Saifuddin, N.; Raziah, A.Z.; Junizah, A.R. Carbon Nanotubes: A Review on Structure and Their Interaction with Proteins. *J. Chem.* **2013**, *2013*, 676815. [CrossRef]
11. Yanar, N.; Yang, E.; Park, H.; Son, M.; Choi, H. Boron Nitride Nanotube (BNNT) Membranes for Energy and Environmental Applications. *Membranes* **2020**, *10*, 430. [CrossRef]
12. Nematollahi, P.; Esrafili, M.D.; Bagheri, A. Functionalization of single-walled (n, 0) carbon and boron nitride nanotubes by carbonyl derivatives (n = 5, 6): A DFT Study. *Can. J. Chem.* **2016**, *94*, 2–21. [CrossRef]

13. Gharbavi, K.; Badehian, H. Structural and electronic properties of armchair (7, 7) carbon nanotubes using DFT. *Comput. Mater. Sci.* **2014**, *82*, 159–164. [CrossRef]
14. Itas, Y.S.; Ndikilar, C.E.; Zangina, T.; Hafeez, H.Y.; Safana, A.A.; Khandaker, M.U.; Ahmad, P.; Abdullahi, I.; Olawumi, B.K.; Babaji, M.A.; et al. Synthesis of Thermally Stable h-BN-CNT Hetero-Structures via Microwave Heating of Ethylene under Nickel, Iron, and Silver Catalysts. *Crystal* **2021**, *11*, 1097. [CrossRef]
15. Ahmadi, S.; Raeisi, M.; Eslami, L.; Rajabpour, A. Thermoelectric Characteristics of Two-Dimensional Structures for Three Different Lattice Compounds of B-C-N and Graphene Counterpart BX (X = P, As, and Sb) Systems. *J. Phys. Chem.* **2021**, *125*, 14525–14537. [CrossRef]
16. Shao, J.; Beaufils, C.; Kolmogorov, A.N. Ab initio engineering of materials with stacked hexagonal tin frameworks. *Sci. Rep.* **2016**, *6*, 28369. [CrossRef]
17. Omidvar, A.; Hadipour, N. Density functional theory studies of carbon nanotube-Graphene nanoribbon hybrids. *J. Iran. Chem. Soc.* **2013**, *10*, 1239–1246. [CrossRef]
18. An, W.; Turner, H. Linking Carbon and Boron-Nitride Nanotubes: Heterojunction Energetics and Band Gap Tuning. *J. Phys. Chem. Lett.* **2010**, *1*, 2269–2273. [CrossRef]
19. Kostoglou, N. Boron Nitride Nanotubes Versus Carbon Nanotubes: A Thermal Stability and Oxidation Behavior Study. *Nanomaterials* **2020**, *10*, 2435. [CrossRef]
20. An, Y.; Sun, Y.; Jiao, J.; Zhang, M.; Wang, K.; Chen, X.; Wu, D.; Wang, T.; Fu, Z.; Jiao, Z. The rectifying effect of heterojunctions composed of carbon and boron nitride nanotubes. *Org. Electron.* **2017**, *50*, 43–47. [CrossRef]
21. El-Barbary, A.A.; Eid, K.M.; Kamel, M.A.; Taha, H.O.; Ismail, G.H. Adsorption of CO, CO_2, NO and NO_2 on Carbon Boron Nitride Hetero Junction: DFT Study. *J. Surf. Eng. Mater. Adv. Technol.* **2015**, *5*, 57979. [CrossRef]
22. Hong-Xia, L.; He-Ming, Z.; Jiu-Xu, S.; Zhi-Yong, Z. Electronic transport properties of an (8, 0) carbon/boron nitride nanotube heterojunction. *Chin. Phys. B* **2010**, *19*, 037104. [CrossRef]
23. Liu, H.; Turner, H. Oxygen Adsorption Characteristics on Hybrid Carbon and Boron-Nitride Nanotubes. *Comput. Chem.* **2014**, *35*, 1058–1063. [CrossRef] [PubMed]
24. Yap, Y.K. *Hetero-Junctions of Boron Nitride and Carbon Nanotubes: Synthesis and Characterization*; Technical Report No. DOE-MTU-ER46294; Michigan Technological University: Houghton, MI, USA, 2013; pp. 1–7.
25. Ahmad, P.; Khandaker, M.U.; Khan, Z.R.; Amin, Y.M. Synthesis of boron nitride nanotubes via chemical vapour deposition: A comprehensive review. *RSC Adv.* **2015**, *5*, 35116–35137. [CrossRef]
26. Nafiu, S.; Apalangya, V.A.; Yaya, A.; Sabi, E.B. Boron Nitride Nanotubes for Curcumin Delivery as an Anticancer Drug: A DFT Investigation. *Appl. Phys.* **2022**, *12*, 879. [CrossRef]
27. Liu, H.; Zhang, H.; Song, J.; Zhang, Z. Electronic structures of an (8, 0) boron nitride/carbon nanotube heterojunction. *J. Semicond.* **2010**, *31*, 013001. [CrossRef]
28. Dhungana, K.; Pati, R. Boron nitride nanotubes for spintronics. *Sensors* **2014**, *14*, 17655–17685. [CrossRef] [PubMed]
29. Qian, L.; Xie, Y.; Zhang, S.; Zhang, J. Band Engineering of Carbon Nanotubes for Device Applications. *Matter* **2020**, *3*, 664–694. [CrossRef]
30. Chen, C.-W.; Lee, M.-H.; Clark, S. Band gap modification of single-walled carbon nanotube and boron nitride nanotube under a transverse electric field. *Nanotechnology* **2004**, *15*, 1837–1843. [CrossRef]
31. Liu, X.; Han, M.; Zhang, X.; Hou, H.; Pang, S.; Wu, Q. Tuning Electronic Structures of BN and C Double-Wall Hetero-Nanotubes. *J. Nanomater.* **2015**, *2015*, 326294. [CrossRef]
32. Akter, N.; Mawardi Ayob, M.T.; Radiman, S.; Khandaker, M.U.; Osman, H.; Alamri, S. Bio-Surfactant Assisted Aqueous Exfoliation of High-Quality Few-Layered Graphene. *Crystals* **2021**, *11*, 944. [CrossRef]
33. Yuan, Y.; Wang, F. A comparison of three DFT exchange-correlation functionals and two basis sets for the prediction of the conformation distribution of hydrated polyglycine. *J. Chem. Phys.* **2021**, *155*, 094104. [CrossRef] [PubMed]
34. Geiger, R.; Zabel, T.; Sigg, H. Group IV direct band gap photonics: Methods, challenges, and opportunities. *Front. Mater.* **2015**, *2*, 52. [CrossRef]
35. Crowley, J.M.; Tahir-Kheli, J., III. Resolution of the Band Gap Prediction Problem for Materials Design. *J. Phys.* **2016**, *7*, 1198–1203. [CrossRef] [PubMed]
36. Zviagin, V.; Grundmann, M.; Schmidt-Grund, R. Impact of Defects on Magnetic Properties of Spinel Zinc Ferrite Thin Films. *Phys. Status Solidi B* **2020**, *257*, 1900630. [CrossRef]
37. Wang, G. First-Principles Studies of Group IV and Group V Related Two Dimensional Materials. Ph.D. Thesis, Michigan Technological University, Houghton, MI, USA, 2016. [CrossRef]
38. Punter, A.; Nava, P.; Carissan, Y. Atomic pseudopotentials for reproducing π-orbital electron behavior in sp^2 carbon atoms. *Int. J. Quantum Chem.* **2019**, *119*, 25914. [CrossRef]
39. Li, R.; Cao, H.; Dong, J. Electronic properties of group-IV monochalcogenide nanoribbons: Studied from first-principles calculations. *Phys. Lett. A* **2017**, *381*, 3747–3753. [CrossRef]
40. Sun, Y.; Nishida, T. *Band Structures of Strained Semiconductors*; Springer: Boston, MA, USA, 2009. [CrossRef]
41. Landau, A.; Khistyaev, K.; Dolgikh, S.; Krylov, A.I. Frozen natural orbitals for ionized states within equation-of-motion coupled-cluster formalism. *J. Chem. Phys.* **2010**, *132*, 014109. [CrossRef]

42. Steeve, C.; Salahub, D.R. Density Functional Theory, Methods, Techniques, and Applications. In *Atomic Clusters and Nanoparticles. Agregats Atomiques et Nanoparticules*; Springer: Berlin/Heidelberg, Germany, 2000; Volume 73, pp. 105–160.
43. Cohen, A.J.; Mori-Sánchez, P.; Yang, W. Insights into Current Limitations of Density Functional Theory. *Science* **2008**, *321*, 792–794. [CrossRef]
44. Garza, J.; Vargas, R.; Nichols, J.; Dixon, D.A. Orbital energy analysis with respect to LDA and self-interaction corrected exchange-only potentials. *J. Chem. Phys.* **2001**, *114*, 639–651. [CrossRef]
45. Giarusso, S.; Gori-Giorgi, P. Exchange-Correlation Energy Densities and Response Potentials: Two Definitions and Analytical Model for the Strong-Coupling Limit of a Stretched Bond. *J. Phys. Chem.* **2020**, *124*, 2473–2482. [CrossRef]
46. Grüning, M.; Gritsenko, O.V.; Baerends, E.J. Exchange-correlation energy and potential as approximate functionals of occupied and virtual Kohn-Sham orbitals: Application to dissociating H_2. *J. Chem. Phys.* **2003**, *118*, 7183. [CrossRef]

crystals

MDPI

Article

Study of the Unstable Rotational Dynamics of a Tor-Fullerene Molecular System

Vladislav Borodin [1], Mikhail Bubenchikov [2], Alexey Bubenchikov [2], Dmitriy Mamontov [2,*], Sergey Azheev [2] and Alexandr Azheev [2]

1 Gazprom Transgaz Tomsk LLC, Tomsk 634050, Russia
2 Department of Mathematics and Mechanics, National Research Tomsk State University, Tomsk 634050, Russia
* Correspondence: orevaore@mail.ru; Tel.: +7-9521526410

Abstract: This work is devoted to modeling the dynamics of large molecules. The key issue in modeling the dynamics of real molecular systems is to correctly represent the temperature of the system using the available theoretical tools. In most works on molecular dynamics, vibrations of atoms inside a molecule are modeled with enviable persistence, which has nothing to do with physical temperature. These vibrations represent the energy internal to the molecule. Therefore, it should not be present in problems in the dynamics of inert molecular systems. In this work, by means of classical mechanics, it is shown that the simplest system containing only three molecular bodies, due to multiple acts of pair interactions of these bodies, reproduces the temperature even in an extremely complex unstable motion of the system. However, at the same time, it is necessary to separate the stochastic part of the movement from the deterministic one. Calculations also show that translational fluctuations in the motion of molecules make the greatest contribution to temperature. The contribution of rotational energy to the total energy of fluctuation motions is small. It follows from these results that the thermal state of the system is determined only by the translational temperature. The latter, in turn, opens up possibilities for a simplified description of many complex systems composed of carbon molecules such as fullerenes and nanotori.

Keywords: molecular dynamics; mathematical modeling; nanomaterials; nanotori; fullerene; CNT; carbon nanotube; C60

Citation: Borodin, V.; Bubenchikov, M.; Bubenchikov, A.; Mamontov, D.; Azheev, S.; Azheev, A. Study of the Unstable Rotational Dynamics of a Tor-Fullerene Molecular System. *Crystals* **2023**, *13*, 181. https://doi.org/10.3390/cryst13020181

Academic Editors: Walid M. Daoush, Fawad Inam, Mostafa Ghasemi Baboli and Maha M. Khayyat

Received: 20 November 2022
Revised: 13 January 2023
Accepted: 14 January 2023
Published: 20 January 2023

1. Introduction

The study of complex movements of carbon molecules in molecular systems is important for the development of nanorobotics. Such equipment can perform a wide range of functions in long-term space flight conditions. This technique is already being used in a number of medical applications. Typical fragments of molecular structures are carbon nanotori and fullerenes since they have a rigid framework structure. Previous reports [1–5] are devoted to the study of the properties of molecular complexes containing standard fullerenes. These studies found that C60 fullerenes can effectively increase the mechanical strength of SWNTs, by acting as a "barrier" to prevent radial deformation and as an inner wall in a double-walled carbon nanotube, as well as the study of the current–voltage characteristics of a system consisting of fullerene and two-carbon nanotubes. Ref. [6] analyzed the interaction of fullerenes with a graphene ribbon accessible to a groove. Based on this interaction, an oscillator was developed. Ref. [7] investigated the transfer of electrons through a carbon molecular transition, a transition from C_{60} molecules attached to a metal concentrate CNT, leading to a coherent regime. It has been shown that the number of contact points between electrodes and a molecule can play an important role in electric transport. Ref. [8] analyzed the features of the interaction of C_{60} with graphite. This article explores the various mechanisms by which C_{60} can be encapsulated in SWNT. Ref. [9], using molecular dynamics simulations, showed that an electric field applied to a @C_{60}

ion inside a water-filled carbon nanotube can pump water with excellent efficiency. A fully controlled nanoelectromechanical device capable of pumping liquids at the nanoscale is proposed. Ref. [10] studied the interaction of a fullerene with a CNT beam. Based on this study, a change in the oscillation frequency is shown depending on the geometric parameters. Ref. [11] developed a multiscale method for predicting the Young's modulus of polymer nanocomposites reinforced with fullerenes (FRPN) and proposed a polymer nanocomposite with a poly(methyl methacrylate) matrix and reinforcing elements in the form of C_{60}. In the study by [12], by analyzing the changes in intermediates and products formed during thermal conversion, the main dependence of the pyrolysis and graphitization of fullerenes on temperature and carbon sources was revealed. It has been found that a higher temperature significantly accelerates the pyrolysis of fullerenes and produces a huge variety of porous carbon products. Ref. [13] studied the possible formation of methane hydrate containing 8, 10, 12, 14, 16, 18, and 20 water molecules into fullerene C_{240} was studied for the first time. After placing a methane molecule in a fullerene containing 20 water molecules, the closed water molecules formed an almost complete dodecahedron shape, and the methane molecule ended up in the center of the dodecahedron. This is the most ordered polygonal structure of methane clathrate, formed into C_{240} fullerene, which is more stable than other clathrate structures. Ref. [14] studied the mechanical and thermodynamic properties of cis-PI-fullerene (C_{60}) composites and achieved the simulation of coarse-grained molecular dynamics (MD) of cis-PI-C_{60} composites with different concentrations of fullerenes. It was found that the density, bulk modulus, thermal expansion, heat capacity and Tg of NR composites increase with increasing C_{60} concentration. The presence of C_{60} led to a slight increase in the interterminal distance and the radius of rotation of the cis-PI chains. The contribution of C_{60} and cis-PI interfacial interactions led to an increase in the bulk moduli of the composites. Ref. [15] studied how the soot core is exposed to polycyclic aromatic hydrocarbons and then grows due to the condensation of polycyclic aromatic hydrocarbons. A nonbonding interaction between polycyclic aromatic hydrocarbons and carbon black has been studied using the distribution of free energy during dimerization and condensation. Ref. [16] are also using molecular dynamics simulations to evaluate the electrical interactions of fullerenes with ions contained in water. The results indicate a smaller chemical shift $\delta(^{13}C)$ and a more intense electronic transition band for fully polarized C_{60} in solution. In [17–21], studies of fullerenes and carbon nanotori by molecular dynamics methods are presented. Extensive results of computational molecular dynamics of various structures were obtained, and interaction potentials and mathematical models were developed.

In the present work, the state of the nanothory system and two fullerenes inside are studied via molecular dynamics methods. It is quite possible to call such a system compact, since molecular objects are together for an arbitrarily long time. The purpose of this work is to define the molecular temperature as an average stochastic characteristic determined by multiple impacts of fullerenes on a nanotorus. The existence of such a temperature indicates the presence of dynamic equilibrium in the system.

2. Computational and Theoretical Models

Models of the dynamics of large molecules are used in the work. This means that individual fragments of the molecular system move as non-deformable molecular shells, interacting with each other and, perhaps, with external electromagnetic fields. At the same time, the interaction of two molecular objects with each other is the total result of the interaction of each atom of one molecular body with all atoms of another. There are three molecular bodies in the case under consideration: a nanotorus and two fullerenes. The rotational motion of each of these bodies can be described using vector equations for a change in the kinetic moment (moment of momentum) for each of these bodies:

$$\frac{d\boldsymbol{K}_1}{dt} = \boldsymbol{L}_1, \ \frac{d\boldsymbol{K}_2}{dt} = \boldsymbol{L}_2, \ \frac{d\boldsymbol{K}_3}{dt} = \boldsymbol{L}_3. \tag{1}$$

where K_1, K_2, K_3 are the kinetic moments of individual bodies; L_1,L_2,L_3—moments of forces acting on this body from the remaining two molecular bodies. In this case, due to the pairing of van der Waals interactions and the absence of external forces:

$$L_1 + L_2 + L_3 = 0. \quad (2)$$

Then, the combination of Equation (1) can be written as follows:

$$\frac{d}{dt}(K_1 + K_2 + K_3) = 0. \quad (3)$$

From (3), we find by integration:

$$K_1 + K_2 + K_3 = C, \quad (4)$$

Here, C is a vector constant. Thus, if no external forces act on a molecular system, then its total angular momentum retains its value both in magnitude and direction.

The angular momentum of the first molecular body K_j is the product of the tensor of inertia $\hat{J}_j$, and the column vector of the instantaneous angular velocity of this body ω_j:

$$K_1 = \hat{J}_j \omega_j. \ (j = 1, 2, 3). \quad (5)$$

Further omitting the index that determines the particular body for the tensor of inertia, the following can be written:

$$\hat{J} = \begin{pmatrix} m\sum(y_i^2 + z_i^2) & -m\sum x_i y_i & -m\sum x_i z_i \\ -m\sum y_i x_i & m\sum(x_i^2 + z_i^2) & -m\sum y_i z_i \\ -m\sum x_i z_i & -m\sum y_i z_i & m\sum(x_i^2 + y_i^2) \end{pmatrix}. \quad (6)$$

where the summation is over the number of atoms that make up the molecule; x_i,y_i,z_i are the coordinates of the i-th atom; m is the mass of the carbon atom. Thus, the components of the inertia tensor of a molecular body are the summation (over the entire number of atoms) of the quadratic functions of the coordinates of the individual atoms that make up the molecule under consideration. The coordinates of all nodes of the molecular body are calculated using the following formula:

$$\frac{dr_i}{dt} = \omega \times r_i + v_c \ \left(i = \overline{1, N}\right). \quad (7)$$

where $r_i = (x_i,y_i,z_i)$; N is the number of atoms that make up the molecular body; ω is the instantaneous angular velocity vector of the considered molecule; v_c is the velocity of the center of mass of this molecule. The last relation expresses the theorem of the addition of velocities in the complex motion of a material point, namely, together with the center of mass and around it. For the velocities of the centers of mass of all molecular bodies participating in the interaction, the equations of motion of their centers of mass are valid:

$$F_1 = M_1 \frac{dv_{c1}}{dt}, \ F_2 = M_2 \frac{dv_{c2}}{dt}, \ F_3 = M_3 \frac{dv_{c3}}{dt}. \quad (8)$$

where M_1 is the nanotorus mass, M_2, M_3 are the fullerene masses, F_1 is the sum of all forces acting on the nanotori atoms from the fullerene atoms; F_2 and F_3 are the total forces of action on the fullerene under consideration from the nanotori and the remaining fullerene. As can be seen from the written relations, the center of mass of the molecule will move at a speed noticeably lower than that of fullerenes since its mass is much greater. In this case, each atom–atom interaction between nodes belonging to different molecular bodies is determined by the gradient of the interaction potential used. The origin of coordinates is chosen at the center of mass of the system. In this case, the components of the tensor of inertia will be calculated by Equation (6) and will be functions of time. The new position of the molecular body in space will be determined by the coordinates of all its points, i.e., coordinates of carbon atoms. If the molecular body is large, then the number of calculated coordinates can be reduced by collecting atoms at the center of mass of some representative fragment. As can be seen from the above description, there is no need to use the Euler angles, as well as the Euler kinematic relations, which have a coordinate singularity. All calculations were carried out according to the Runge–Kutta scheme [22] of the fourth order of accuracy with a constant time step $\Delta t = 10^{-6}$ ns. The accuracy of the calculations was checked by performing the balance of the total energy of the system and amounted to 10^{-8} relative units. The calculated kinetic energy was related to the initial potential energy of the interaction of molecular bodies. The LJ potential of atom–atom interactions was used in the calculations. This is a classical potential: attraction–repulsion, which has two parameters of interaction: ε is the depth of the potential well, and σ is the radius of influence of the interaction. Leaving σ unchanged (0.34 nm for carbon atoms), we select the depth of the potential well in such a way that the vibration energy of the bodies participating in the movement corresponds to a certain temperature, in this case, room temperature T = 300 K (ε/k = 5.1 K, where k is Boltzmann constant). The vibrations of carbon atoms in the molecular bodies under consideration must correspond to this temperature. In this case, however, due to the small mass of atoms and strong C-C bonds, they have a higher frequency and a significantly lower amplitude. Their inclusion in the consideration does not affect the nature of movements on larger scales. Therefore, they remain small-scale and high-frequency, forming the background against which larger-scale molecular events unfold.

3. Results

For solving the system of such equations, step-by-step integration schemes of the Runge–Kutta class of the fourth order with a constant time step $\Delta t = 10^{-7}$ ns are used. All calculations were carried out using the software developed by the authors according to the scheme proposed in this work for determining the rotations of molecular bodies in space.

Figure 1 presents the case of rotation of the torus with an initial angular velocity of $\omega_x^0 = 150\ ns^{-1}$, $\omega_y^0 = \omega_z^0 = 0$. In this case, the fullerenes are located diametrically opposite on the (oy) axis and are fixed on the axial circumference of the torus. The red line in this figure is the trajectory of one of the peripheral carbon atoms that does not lie on the (ox) axis and has a y-coordinate value close to the outer radius of the torus. This line demonstrates somersaults of the considered molecular structure during its rotation around an axis with an intermediate moment of inertia (oy axis), i.e., demonstrates the instability of Louis Poinsot [23–25]. The (oz) axis is directed perpendicular to the initial position of the torus plane. In what follows, the motion of the molecular structure under consideration is analyzed in the absolute frame of reference thus introduced.

In addition, in this article, the rotation of a tor-fullerene molecular system is considered. Thus, we compare the case of fullerenes having internal freedom with the case of fullerenes maximally separated and fixed inside the torus. The nanotorus is composed of 2000 carbon atoms and has an inner radius R_1 = 0.3 nm, R_2 = 1.9 nm (Figure 2). It contains standard C_{60} fullerenes. In all variants, the fullerenes are maximally spaced at the initial moment, and the rotation is carried out around an axis lying in the plane of the torus and passing perpendicular to the straight line connecting the centers of mass of the fullerenes. In the case of loose fullerenes, the rotation of the system around an axis with an intermediate moment of inertia forms orbital displacements of fullerenes along the axis of the nanotorus channel.

The data are reduced to a moving frame of reference with coordinate axes lying in the rotating plane of the torus. It turned out that somersaults of the torus induce reciprocating movements of fullerenes along the curvilinear axis of the torus. Moreover, the center of such oscillations shifts in a certain direction. The direction of axial displacement of buckyballs is determined by the right hand rule. If the vector of the initial angular velocity of the molecular structure enters the palm of the right hand, and the moved thumb is perpendicular to the plane of the torus and points to the vector of its angular rotation, and then the index finger shows in which direction the fullerenes move. The red line in the center of Figure 2 shows the trajectory of the center of mass of the nanotorus in this case. The center of mass of the torus is displaced due to the movement of fullerenes; however, the center of mass of the entire torus–fullerene structure remains immobile. The case of the rotation of such a system, which has some freedom of movement for fullerenes inside the torus, demonstrates one of the options for the transition of the rotational motion of the system into the translational motion of fullerenes along the circle, which is the axial line of the torus channel. It should be noted that in this case, fullerenes receive rotations in intensity significantly exceeding their initial values. These metamorphoses are associated with a decrease in the rotational energy of the frame body, which has a significantly larger mass than fullerenes. These peculiarities of motion are associated with a decrease in the rotational energy of the frame body, which has a much larger mass than fullerenes. The nature of the angular vibrations of fullerenes is shown in Figure 3.

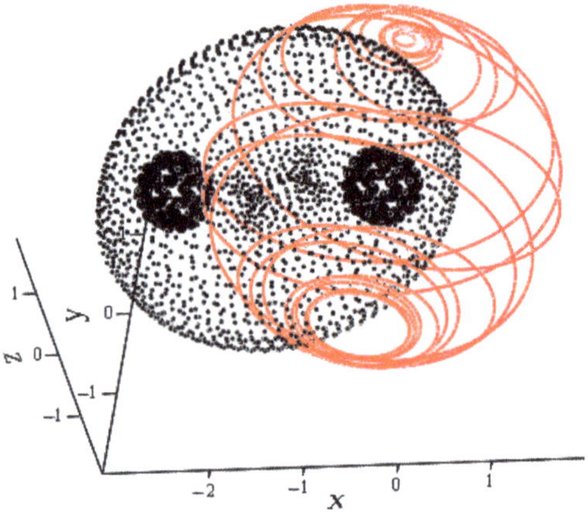

Figure 1. Unstable rotation of a tor-fullerene molecular structure. The red line shows the trajectory of one of the nanotorus atoms.

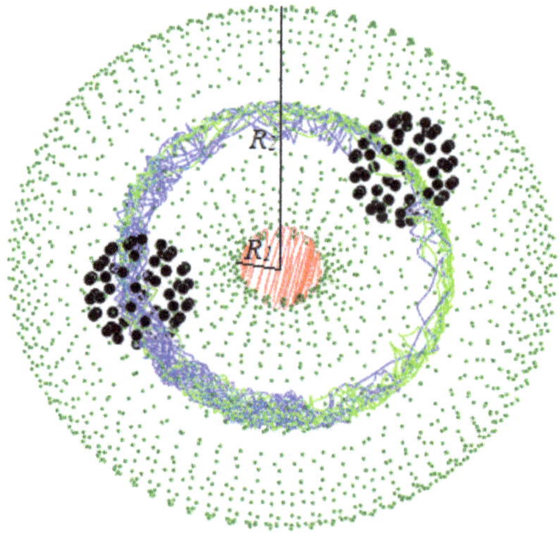

Figure 2. Orbital motion of fullerenes. In the center of the figure, the trajectory of the center of mass of the nanotorus; inside the torus, the trajectory of the centers of mass of fullerenes.

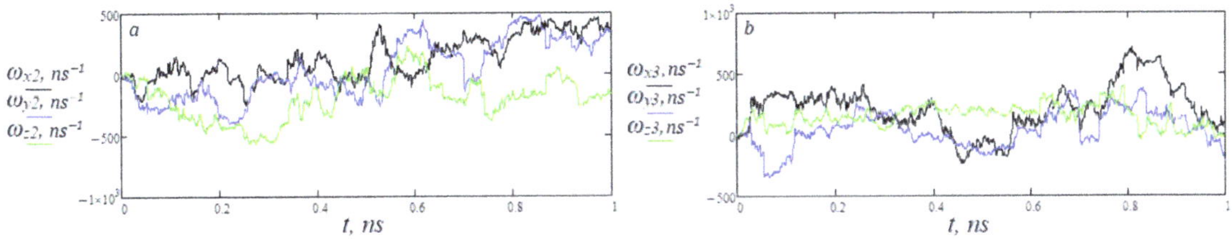

Figure 3. Projections of the angular velocities of fullerenes. (**a**)—the case of "frozen" fullerenes. (**b**)—case of free fullerenes.

Figure 3 shows that the frequencies of angular vibrations of fullerenes are at least 3 times higher than the main frequency of rotation of the body part of the molecular struc-ture. Next, we compare two variants of calculations of the motion of the molecular struc-ture under consideration. Figure 3a refers to the case of pinned fullerenes. Figure 3b variant, fullerenes can move freely in the accessible zones of the inner space of the torus.

Figures 4–6 show the projections of the angular velocities of the torus in these cases. It can be seen that the rotation of a single structure is periodic at the beginning of the movement. In the second case, some quasi-periodic motion is established on a time interval of 1 ns. Figure 4 shows what part of the initial rotational energy of the torus is converted into the kinetic energy of fullerenes. As can be seen from Figures 4 and 6, in the case of "frozen" fullerenes, one can clearly distinguish bi-oscillations, which are determined by the presence of two frequencies—the initial frequency of rotations of the system and the frequency of somersaults by V. Dzhanibekov [26].

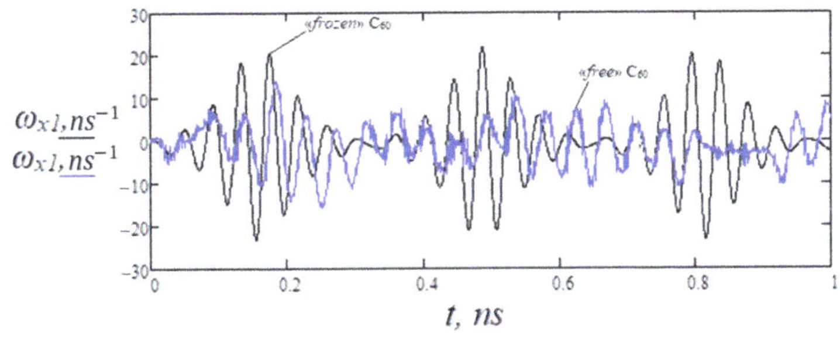

Figure 4. Comparison of rotations of two systems with free and "frozen" fullerenes.

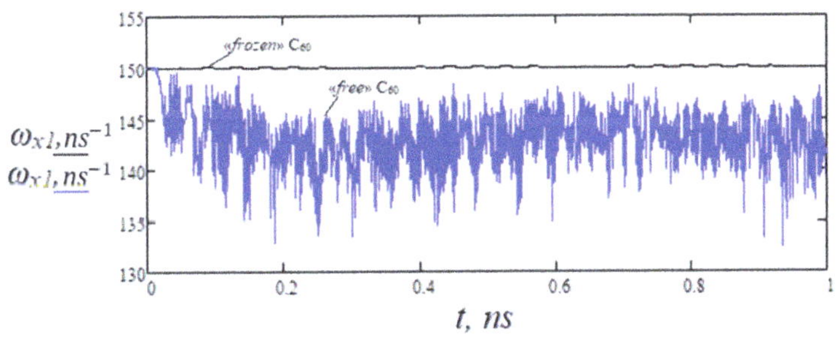

Figure 5. Change in time of the y-component of the angular velocity of the nanotorus.

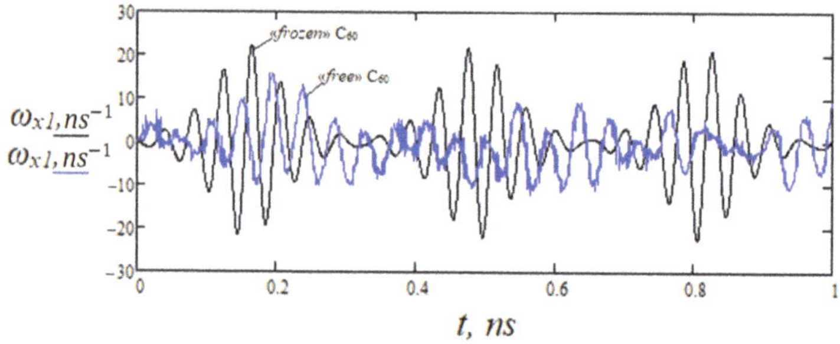

Figure 6. Distribution of the *z*-component of the angular velocity of the nanotorus.

Figure 7 shows the total angular frequency modulus of the nanotorus with "frozen" fullerene. The highest peaks on this graph show "Dzhanibekov flips", and the time between peaks is the period with which these somersaults are made.

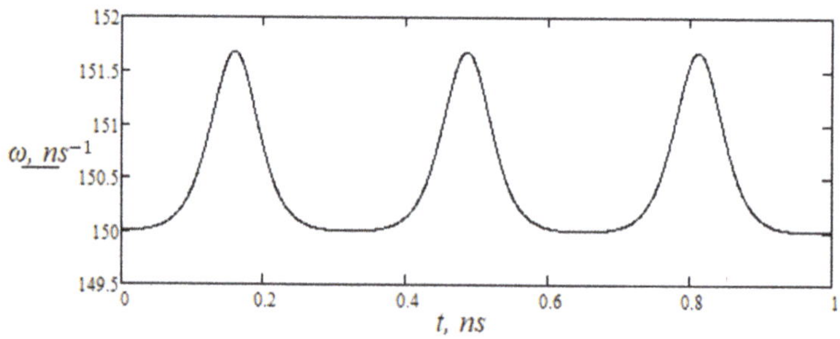

Figure 7. Module of angular velocity of nanotorus with "frozen" fullerene.

Since the inertial rotation of the molecular system is considered, the vector of the angular momentum of the entire system $\boldsymbol{K}$ remains constant during the entire time of motion. There is a non-moving plane perpendicular to this vector, where in the case of one body, the vector of the instantaneous angular velocity of this body will draw a complete picture of the motion instability. This plane is called the Poinsot plane, and the trajectory of the end of the instantaneous angular velocity vector on this plane is called the herpoloid [23]. In the case of several bodies (a torus plus two fullerenes), the Poinsot plane also exists, but the end of the angular velocity vector of any of the existing bodies will not slide along this plane. Nevertheless, the behavior of the projection onto this plane of the vector of the instantaneous angular velocity of the frame body is of interest, i.e., the molecular torus, as a body having a significantly larger mass in comparison with other structural elements.

Figure 8 shows the trajectory of the end of the instantaneous angular velocity vector of a nanotori with fullerenes "frozen" into it. As can be seen from the presented figure, the herpoid in this case is a fairly smooth curve. The general view of the drawing resembles a bud of an unopened flower. The herpoid of a single body with different values of the axial moments of inertia is always a petal structure. It is just that with such a slight difference in the moments of inertia, as in the case under consideration, the petals in the bud are too wide and they are tightly packed. With a more significant difference in the axial moments of inertia, a chamomile herpoid design can be obtained. In any case, if the molecular body is uniform, then the end of the vector of its instantaneous angular velocity moves step by step in one direction along the herpoloid.

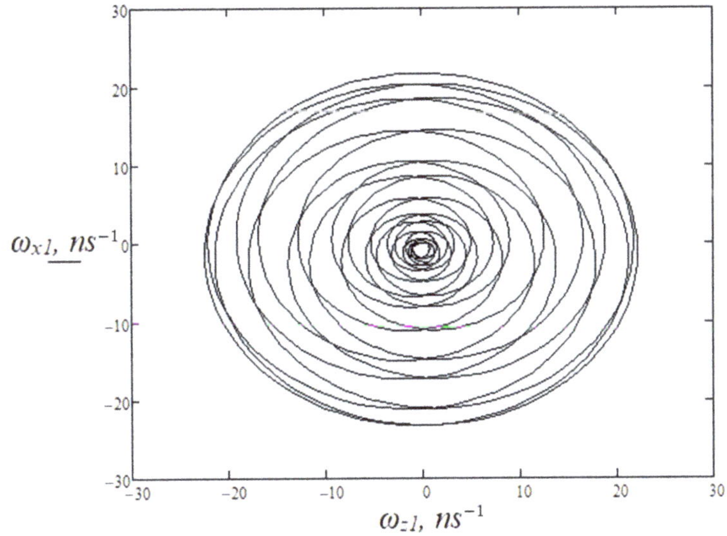

Figure 8. Herpoloid torus with "frozen" fullerenes.

Figure 9 shows what happens to the herpolody when fullerenes are able to move inside the torus space. The motion along the herpoid takes on a reciprocating character, although globally, it remains unidirectional. The kinks on the herpoid are associated with the reciprocating motion of fullerenes inside the torus.

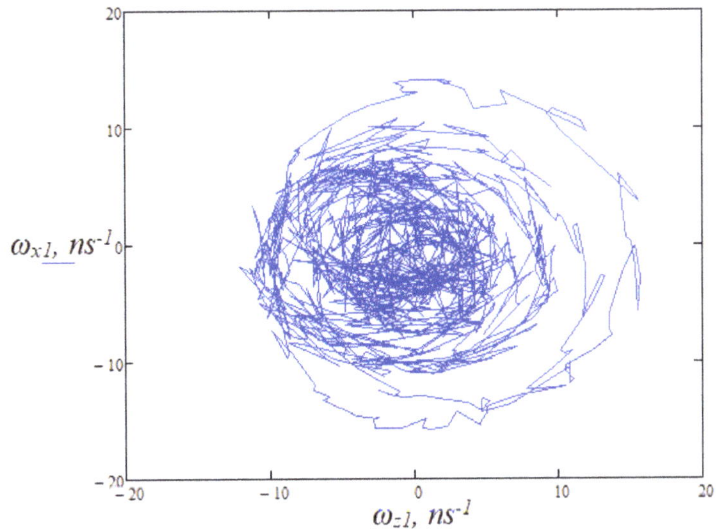

Figure 9. Herpoid composite molecular structure with freely moving fullerenes.

4. Discussion

Despite the fact that the case described by Louis Poinsot is an example of the instability of motion, it is characterized by a regular rotation of a molecular object with strictly periodic flips around the axis with a minimum moment of inertia. Such movement can rightly be attributed to the class of deterministic processes. Calculations show that the interactions of two molecular objects already lead to the appearance of irregular oscillations in this simplest system. All this is a sign of the manifestation of stochastic properties. The two fullerenes considered in this paper inside the nanotorus have random angular rotations, as well as random approaches of fullerenes to the wall. However, under the conditions of the considered problem, the transversal displacements of fullerenes inside the torus turned out to be deterministic. To determine the direction of these movements, the authors adapted the right hand rule. It is known that somersaults of an object rotating by inertia have a certain direction of flips. These flips generate displacements of fullerenes in a strictly defined direction along the axial circumference of the torus. However, these movements are defined only in a global sense, since they have a local reciprocating character. Thus, the system demonstrates stochastic properties all the time of movement. Even fullerenes of small mass create disturbances that greatly loosen the motion of the nanotorus. In this regard, the instantaneous axis of rotation of the nanotorus ceases to smoothly and monotonously change its direction in space. The motion of the entire system becomes largely stochastic. However, it is still possible to single out a deterministic part of it. In this regard, the method of comparing the main variant with the motion of a torus with "frozen" fullerenes turned out to be useful. This case made it possible to isolate the second frequency in the oscillations of the nanotorus due to its flips around the axis with the minimum value of the moment of inertia. Thus, the temperature of a system consisting of a small number of bodies is considered from a not quite classical angle. The molecular dynamics of a very interesting system consisting of fullerenes were also considered. Despite the fact that both nanotori and fullerenes themselves are of great scientific interest and are being studied quite closely [26,27], and similar container-like nanostructures are already proposed for use [28,29], even the theoretical side of these issues has not been fully studied. In this case, any of the found effects or mechanisms can form the basis of various technologies.

5. Conclusions

The interaction of a large nanotorus molecule with a standard C60 fullerene, which makes up only 3% of the mass of the body of the torus, leads to the appearance of thermal motions of both the nanotorus itself and the fullerene contained in it. Moreover, the average values of the energies of random displacements distributed over the six degrees of freedom of each of the molecular bodies are extremely close. This indicates a rapidly emerging equilibrium in the system. In the case of a nanotorus plus two fullerenes considered here, due to multiple impacts on the inner wall of the nanotorus, the equality of the energies of rotational fluctuation displacements of fullerenes with the energy of their translational fluctuations occurs, which ultimately ensures the equality of rotational and vibrational temperatures. The foregoing makes it possible to conclude that there is a single temperature in the system due to the movement of molecular bodies.

Author Contributions: Data curation, V.B.; funding acquisition, M.B.; investigation, A.B.; methodology, M.B., A.B., S.A. and A.A.; project administration, M.B.; resources, M.B.; software, D.M.; supervision, M.B.; validation, A.B.; visualization, D.M.; writing—original draft, D.M. and A.A.; writing—review and editing, A.B. and S.A. All authors have read and agreed to the published version of the manuscript.

Funding: This research was funded by presidential grant MD 4273.2022.1.1.

Data Availability Statement: Not applicable.

Conflicts of Interest: The authors declare no conflict of interest.

References

1. Wu, C.-D.; Fang, T.-H.; Chan, C.-Y. A molecular dynamics simulation of the mechanical characteristics of a C60-filled carbon nanotube under nanoindentation using various carbon nanotube tips. *Carbon* **2011**, *49*, 2053–2061. [CrossRef]
2. Tu, G.-Y.; Guo, D.-P.; Li, B.-R.; Zhang, H.-L. Transport properties and mechanism of C60 coupled to carbon nanotube electrode. *Phys. B Condens. Matter* **2011**, *406*, 2138–2142. [CrossRef]
3. Kang, J.W.; Hwang, H.J. Comparison of C60 encapsulations into carbon and boron nitride nanotubes. *J. Phys. Condens. Matter* **2004**, *16*, 3901. [CrossRef]
4. Shokri, A.; Nikzad, S. Electrical transport properties of a CNT/C60/CNT hybrid junction with closed end CNT leads using Green's function method. *Eur. Phys. J. B* **2013**, *86*, 22. [CrossRef]
5. Zhou, L.; Zhu, B.E.; Pan, Z.Y.; Wang, Y.X.; Zhu, J. Reduction of the buckling strength of carbon nanotubes resulting from encapsulation of C60 fullerenes. *Nanotechnology* **2007**, *18*, 275709. [CrossRef]
6. Kang, J.W.; Lee, K.W. Molecular dynamics study on the C60 oscillator in a graphene nanoribbon trench. *J. Korean Phys. Soc.* **2014**, *65*, 185–189. [CrossRef]
7. Shokri, A.A.; Nikzad, S. Topology effects of interface and gate voltage on electrical transport through the CNT/C60/CNT junction using the Green's function method. *J. Appl. Phys.* **2011**, *110*, 024303. [CrossRef]
8. Ulbricht, H.; Moos, G.; Hertel, T. Interaction ofC60with Carbon Nanotubes and Graphite. *Phys. Rev. Lett.* **2003**, *90*, 095501. [CrossRef]
9. Farimani, A.B.; Heiranian, M.; Aluru, N.R. Nano-electro-mechanical pump: Giant pumping of water in carbon nanotubes. *Sci. Rep.* **2016**, *6*, 26211. [CrossRef]
10. Ansari, R.; Sadeghi, F.; Alipour, A. Oscillation of C60 Fullerene in Carbon Nanotube Bundles. *J. Vib. Acoust.* **2013**, *135*, 051009. [CrossRef]
11. Izadi, R.; Nayebi, A.; Ghavanloo, E. Combined molecular dynamics–micromechanics methods to predict Young's modulus of fullerene-reinforced polymer composites. *Eur. Phys. J. Plus* **2021**, *136*, 816. [CrossRef]
12. Li, S.; Bian, F.; Meng, X.; Zhai, D.; Yang, H.; Qin, G. Ring structure characterization of nanoporous carbon materials prepared by thermal conversion of fullerenes: Insights from ReaxFF molecular dynamics simulations. *Carbon* **2021**, *189*, 484–492. [CrossRef]
13. Abbaspour, M.; Akbarzadeh, H.; Salemi, S.; Tahami, S.F. Formation of methane clathrates into fullerene: A molecular dynamics study. *J. Mol. Liq.* **2022**, *367*, 120587. [CrossRef]
14. Kitjanon, J.; Khuntawee, W.; Phongphanphanee, S.; Sutthibutpong, T.; Chattham, N.; Karttunen, M.; Wong-Ekkabut, J. Nanocomposite of Fullerenes and Natural Rubbers: MARTINI Force Field Molecular Dynamics Simulations. *Polymers* **2021**, *13*, 4044. [CrossRef] [PubMed]
15. Yuan, H.; Kong, W.; Xia, J. Steered molecular dynamics and stability analysis on PAH dimerisation and condensation on fullerene and soot surfaces. *Phys. Chem. Chem. Phys.* **2021**, *23*, 19590–19601. [CrossRef] [PubMed]
16. Cardoso, W.B.; Colherinhas, G. Fullerene C60 spectroscopy in [BMIM][PF6] ionic liquid: Molecular dynamics study using polarization effects. *J. Mol. Struct.* **2021**, *1250*, 131887. [CrossRef]
17. Lun-Fu, A.V.; Bubenchikov, A.M.; Ovchinnikov, V.A. Numerical study of rotational dynamics of carbon nanotori during intermolecular interaction. *J. Phys. Conf. Ser.* **2022**, *2211*, 012004. [CrossRef]
18. Lun-Fu, A.; Bubenchikov, M.; Bubenchikov, A.; Mamontov, D.; Borodin, V. Interaction of molecular tori in columnar structures. *J. Phys. Condens. Matter* **2022**, *34*, 125101. [CrossRef]
19. Bubenchikov, M.; Mamontov, D.; Ovchinnikov, V. Studying the possibility of using fullerenes inside carbon nanotubes as a molecular engine. *J. Phys. Conf. Ser.* **2022**, *2211*, 012005. [CrossRef]
20. Bubenchikov, A.M.; Bubenchikov, M.A.; Mamontov, D.V.; Chelnokova, A.S.; Chumakova, S.P. Movement of fullerenes and their dimers inside carbon nanotubes. *Full- Nanotub. Carbon Nanostruct.* **2021**, *29*, 803–809. [CrossRef]
21. Bubenchikov, A.M.; Bubenchikov, M.A.; Mamontov, D.V.; Kaparulin, D.S.; Lun-Fu, A.V. Dynamic state of columnar structures formed on the basis of carbon nanotori. *Full- Nanotub. Carbon Nanostruct.* **2021**, *29*, 825–831. [CrossRef]
22. Runge, C. Ueber die numerische Auflösung von Differentialgleichungen. *Math. Ann.* **1895**, *46*, 167–178. [CrossRef]
23. Poinsot, L. *Theéorie Nouvelle De La Rotation Des Corps*; Bachelier: Paris, France, 1834; Volume 1834, pp. 1777–1859.
24. Levi, M. Classical Mechanics with Calculus of Variations, and Optimal Control: An Intuitive Introduction. *Am. Math. Soc.* **2014**, *69*, 1470414449. [CrossRef]
25. Trivisonno, N.; Garelli, L.; Storti, M. The Tennis Racket Theorem, an Analysis of the Intermediate Axis Theorem. *Mecánica Comput.* **2021**, *38*, 1341–1353.
26. Sarapat, P.; Hill, J.M.; Baowan, D. A Review of Geometry, Construction and Modelling for Carbon Nanotori. *Appl. Sci.* **2019**, *9*, 2301. [CrossRef]
27. Shi, J.; Wang, B.; Wang, L.; Lu, T.; Fu, Y.; Zhang, H.; Zhang, Z. Fullerene (C60)-based tumor-targeting nanoparticles with "off-on" state for enhanced treatment of cancer. *J. Control Release* **2016**, *235*, 245–258. [CrossRef] [PubMed]

28. Chernozatonskii, L.A.; Artyukh, A.A.; Demin, V. Quasi-one-dimensional fullerene-nanotube composites: Structure, formation energetics, and electronic properties. *JETP Lett.* **2013**, *97*, 113–119. [CrossRef]
29. Singhai, N.J.; Ramteke, S. Functionalized Carbon Nanotubes: Emerging Applications in the Diverse Biomedical Arena. *Curr. Nanosci.* **2020**, *16*, 170–186. [CrossRef]

crystals

MDPI

Article

Fullerene in a Magnetic Field

Alexandr Lun-Fu [1], Vladislav Borodin [1], Mikhail Bubenchikov [2], Alexey Bubenchikov [2] and Dmitriy Mamontov [2,*]

1 Gazprom Transgaz Tomsk LLC, 634029 Tomsk, Russia; a.lunfu@gtt.gazprom.ru (A.L.-F.); v.borodin@gtt.gazprom.ru (V.B.)
2 Department of Mathematics and Mechanics, National Research Tomsk State University, 634050 Tomsk, Russia; michael121@mail.ru (M.B.); bubenchikov_am@mail.ru (A.B.)
* Correspondence: orevaore@mail.ru; Tel.: +7-952-152-6410

Abstract: The manuscript presents a trajectory method for describing the rotations of surface crystals such as fullerenes, nanotubes, and nanotori. This method does not require the implementation of successive rotations of the considered molecular structures around the axes of the selected basis. Therefore, it is free from the shortcomings of the approaches of Euler and Hamilton. On its basis, an efficient algorithm for calculating the motions of a magneto-susceptible fullerene in an alternating magnetic field has been developed. The nature of rotation of fullerenes in fields of various configurations has been studied.

Keywords: mathematical modeling; molecular dynamics; nanomaterials

Citation: Lun-Fu, A.; Borodin, V.; Bubenchikov, M.; Bubenchikov, A.; Mamontov, D. Fullerene in a Magnetic Field. *Crystals* **2022**, *12*, 521. https://doi.org/10.3390/cryst12040521

Academic Editors: Conrad Becker and Marilena Carbone

Received: 21 February 2022
Accepted: 5 April 2022
Published: 8 April 2022

1. Introduction

To date, a whole class of surface crystals closed on themselves, called buckyballs, has been synthesized. Many isomers of fullerenes have been studied. This class will noticeably expand if atoms of metals, nitrogen, hydrogen, etc., are added to carbon structures [1,2]. The resulting molecular systems have different properties. Moreover, these properties can differ greatly even due to minor changes—defects, temperature, chemical factors, and exposure to fields [3–6]. In order to act on a molecular fullerene structure with a magnetic field, it is necessary that it contains iron atoms or other ferromagnetic material. The magnetic fullerene system can practically be obtained both as a result of an exoendral arrangement of the K_3C_{60} type and with an endohedral arrangement of iron, i.e., Fe@C_{60}. At the same time, none of the species has yet been prepared, but the work on the synthesis of magnetic materials is quite active [7–10]. In this work, we will proceed from the simplest design of Fe@C_{60}. This makes it possible to spin fullerenes by an alternating field. Of particular interest are fullerites created on the basis of magnetically susceptible fullerenes. In these molecular crystals, due to the high degree of symmetry of the location of the crystal lattice sites and the symmetry of the molecule itself, fullerenes can be unwound without changing their vibrational movements around their centers of mass (without changing the crystal temperature). There are potential opportunities for such unwinding since, in an ideal C_{60}-based fullerite, large-amplitude angular vibrations of molecules are realized. According to [11], their frequency is 34 GHz. In this case, we can talk about the independence of rotational and translational motions, and the material will not be destroyed even with an intensive unwinding of the crystal lattice nodes. Moreover, studies [12–17] have shown that C_{60} fullerenes have a gyroscopic effect, which means that the material becomes more resistant to deformation. Thus, it is possible to accumulate energy on rotational degrees of freedom and thereby control the hardness of molecular crystals. The biomedical focus of articles on molecular robots has a very important aspect related to the release of drugs from the carbon container during their targeted delivery to the affected areas. One of the ways of such release is the unwinding of fullerenes or their dimers [18] by an external alternating magnetic field. In this theoretical work, we wanted to answer the question of the possibility of spinning up a magneto-susceptible fullerene under the action of an alternating magnetic field. By implementing the developed numerical model, we found the answer to this question.

2. Materials and Methods

We will consider the problem of the rotation of a supermolecule that has its own magnetic moment and is in an alternating magnetic field $\boldsymbol{B} = \boldsymbol{B}(t)$:

$$B_x = B_x(t),\ B_y = B_y(t),\ B_z = B_z(t). \tag{1}$$

Here B_x, B_y, B_z are the projections of the magnetic induction vector on the axes of the absolute basis, the center of which we place at the center of mass of the fullerene.

The specific configuration of the magnetic field, i.e., the distribution of its intensity by components, will be presented below in the calculation results section. Now let us consider a scheme for solving the problem in the case of an arbitrary law of changes in the field components with time. We will use the equations of rotational motion written in the absolute basis (theorem on the change in the angular momentum):

$$\frac{d\boldsymbol{K}}{dt} = \boldsymbol{M}. \tag{2}$$

The projections K_x, K_y, K_z of the vector $\boldsymbol{K}$ are defined by the following relationships:

$$\begin{aligned} K_x &= A\omega_x + F\omega_y + E\omega_z, \\ K_y &= F\omega_x + B\omega_y + D\omega_z, \\ K_z &= E\omega_x + D\omega_y + C\omega_z. \end{aligned} \tag{3}$$

Here A, B, C are axial, and D, E, F are centrifugal moments of inertia, which are determined by known formulas:

$$A = m\sum\left(y_i^2 + z_i^2\right), B = m\sum\left(z_i^2 + x_i^2\right), C = m\sum\left(x_i^2 + y_i^2\right). \tag{4}$$

$$D = -m\sum y_i z_i,\ E = -m\sum z_i x_i,\ F = -m\sum x_i y_i. \tag{5}$$

The moment of magnetic action $\boldsymbol{M}$ on the fullerene is the vector product of its own magnetic moment and the magnitude of the magnetic field strength $\boldsymbol{B}$:

$$\begin{aligned} \boldsymbol{M} = \boldsymbol{\mu} \times \boldsymbol{B} = \mu \begin{vmatrix} \boldsymbol{i} & \boldsymbol{j} & \boldsymbol{k} \\ \alpha & \beta & \gamma \\ B_x & B_y & B_z \end{vmatrix} = \\ = \mu(\beta B_z - \gamma B_y)\boldsymbol{i} + \mu(\gamma B_x - \alpha B_z)\boldsymbol{j} + \mu(\alpha B_y - \beta B_x)\boldsymbol{k}. \end{aligned} \tag{6}$$

Here $\boldsymbol{i}, \boldsymbol{j}, \boldsymbol{k}$ are absolute basis vectors; α, β, γ coordinates of the unit vector, coinciding in direction with the vector of the intrinsic magnetic moment $\boldsymbol{\mu}$. If the magnetic moment is directed along the radius of the fullerene, then the coordinates of the unit vector can be found from the coordinates of any atom of the supermolecule as follows:

$$\alpha = \frac{x_k}{r_k};\ \beta = \frac{y_k}{r_k};\ \gamma = \frac{z_k}{r_k}, \tag{7}$$

where $r_k = \sqrt{x_k^2 + y_k^2 + z_k^2}$. As can be seen from the written representations in the equation for changing the angular momentum, the differential sign includes the following functions of time: $\omega_x(t),\ \omega_y(t),\ \omega_z(t)$, and also $x_i(t),\ y_i(t),\ z_i(t)\ \left(i = \overline{1, N}\right)$. All of these features are currently unknown. However, there are too many of them to solve the problem only on the basis of three scalar Equation (2). Euler's theorem is known that a complex spatial rotation of a body can always be represented as a rotation around some axis with an instantaneous angular velocity. Using this theorem in the case of rotational motion of fullerene around its center of mass, we can write:

$$\frac{d\mathbf{r}_i}{dt} = \boldsymbol{\omega} \times \mathbf{r}_i \ \left(i = \overline{1, N}\right). \tag{8}$$

Here $\mathbf{r}_i = (x_i, y_i, z_i)$, $\boldsymbol{\omega} = (\omega_x, \omega_y, \omega_z)$. Moreover, the coordinates of these vectors are presented in an absolute basis. Since we consider the framework structure of the fullerene to be non-deformable, the vector $\boldsymbol{\omega}$ will be the same for all fullerene nodes, i.e., will not depend on the number "i". Thus, the equations for the coordinates of fullerene nodes $x_i(t)$, $y_i(t)$, $z_i(t)$ $\left(i = \overline{1, N}\right)$ are kinematic relations (8). As a result, we have obtained a closed system of scalar equations for finding all unknown time functions that appear under the differential sign on the left side of (2). Differential Equations (2) and (8) are integrated under the following initial conditions:

$$\begin{aligned} t = 0, \quad & K_x = A^0\omega_x^0 + F^0\omega_y^0 + E^0\omega_z^0, \\ & K_y = F^0\omega_x^0 + B^0\omega_y^0 + D^0\omega_z^0, \\ & K_z = E^0\omega_x^0 + D^0\omega_y^0 + C^0\omega_z^0, \\ x_i = x_i^0, \quad & y_i = y_i^0,\ z_i = z_i^0 \ \left(i = \overline{1, N}\right). \end{aligned} \tag{9}$$

Here, the index zero at the top marks the initial values of all considered quantities. Thus, relations (2)–(9) determine the formulation of the problem of fullerene rotations in alternating magnetic fields. One-step methods of a high order of accuracy, to which it belongs, and the standard Runge–Kutta scheme of the fourth order of accuracy use the idea of recalculating the desired values at intermediate positions of a single time step. Therefore, all calculated values are determined both on time layers and at intermediate positions between layers. However, it follows from the above relations that the components of the instantaneous angular velocity vector of a molecular object are found each time from the solution of an algebraic system of equations using new values of the angular momentum projections on the right-hand sides of these algebraic equations.

The presented method for calculating the rotation of bodies does not use the Euler angles connecting the relative positions of the absolute and local bases. Therefore, when implementing it, there is no need to use the kinematic Euler relations, which have a coordinate singularity at nutation angles that are multiples of π. In this regard, studies carried out on the basis of the presented mathematical model have no restrictions on nutation angles. Moreover, we do not use Hamiltonian quaternions. Therefore, in this study, we approve the third method of specifying the rotation of bodies in space, associated with the calculation of the projections of the angular velocity of the body on the axes of the absolute basis and the determination of the instantaneous position of the body by a set of coordinates of representative points of the body. A distinctive feature of calculations of body rotations in the absolute basis is the absence of successive rotations at the level of elementary angles $\omega_x \frac{\Delta t}{2}$, $\omega_y \frac{\Delta t}{2}$, $\omega_z \frac{\Delta t}{2}$. This is achieved by simultaneously finding the projections of angular velocities as a result of solving a system of algebraic equations according to Cramer's rule.

Conservation of Total Energy in the System

Let a fullerene, having its own magnetic moment, be placed in a constant magnetic field B_0 = const. If these vectors do not match, then the object, which has its own magnetic moment, will oscillate endlessly. We will assume that at the initial moment of time, the arrow is deviated from the direction of the magnetic field by an angle $\varphi - \varphi_0$. The total energy balance of a moving magnetic needle (magnetic pendulum) will be determined by the following relationship:

$$\frac{J\omega^2}{2} + E_M^0 - \Delta E_M = E_M^0 = const. \tag{10}$$

Here $\omega = \sqrt{\omega_x^2 + \omega_y^2 + \omega_z^2}$ is the local value of the angular velocity of the fullerene molecule; $J = A\alpha_1{}^2 + B\beta_1{}^2 + C\gamma_1{}^2 + 2D\beta_1\gamma_1 + 2E\gamma_1\alpha_1 + 2F\alpha_1\beta_1$—the magnitude of the moment of inertia of the supermolecule relative to the instantaneous axis of rotation; $\alpha_1, \beta_1, \gamma_1$—direction cosines of this axis with respect to the axes of the absolute basis.

The characteristic energy in the magnetic pendulum problem is E_M^0. This value can be calculated analytically, then the relative error of the fulfillment of relation (10) will be as follows:

$$\varepsilon = \frac{J\frac{\omega^2}{2} - \Delta E_M}{E_M^0}. \tag{11}$$

The magnitude of the magnetic moment is:

$$M = \mu B_0 \sin\varphi. \tag{12}$$

The characteristic potential energy of the magnetic field and the change in this energy when the arrow turns as an angle of $\varphi - \varphi_0$ are determined by the relations:

$$E_M^0 = \mu B_0 \int_0^{\varphi_0} \sin\varphi d\varphi = -\mu_0 B_0 \cos\varphi \Big|_0^{\varphi_0} = \mu B_0 (1 - \cos\varphi_0), \tag{13}$$

$$\Delta E_M = \mu B_0 (\cos\varphi - \cos\varphi_0). \tag{14}$$

3. Results of Calculations

Two field configurations were considered. One of these configurations:

$$B_x = B_0 \sin(ft),\ B_y = 0,\ B_z = 0. \tag{15}$$

The other is a rotating field:

$$B_x = B_0 \sin(ft),\ B_y = B_0 \cos(ft),\ B_z = 0. \tag{16}$$

It was assumed in the calculations that the value of the intrinsic magnetic moment of the fullerene is equal to the magnetic moment of the iron atom. We consider this atom to be intercalated into the framework structure of fullerene. The oscillation amplitude of the alternating magnetic field was taken equal to $B_0 = 1\ T$.

According to the presented mathematical model, it is possible to determine the projections of the instantaneous angular velocity vector on the axes of the absolute basis, as well as the trajectories of the nodes of the molecular structure. Cases with different frequencies of field change were calculated. Two typical cases f = 100 GHz and f = 1000 GHz, are shown in Figures 1–4. In the case of a unidirectional field, the fullerene moves parallel to the plane, passing through the initial position of the magnetic field vectors and intrinsic magnetic moment.

The Figure 1 refers to the rotation of a magneto-susceptible fullerene in a unidirectional magnetic field with a frequency f = 100 GHz.

The Figure 2 refers to the rotation of a fullerene intercalated with iron in a unidirectional magnetic field with a frequency f = 1000 GHZ.

The Figure 3 refers to the spatial rotations of fullerene in a rotating magnetic field f = 100 GHz.

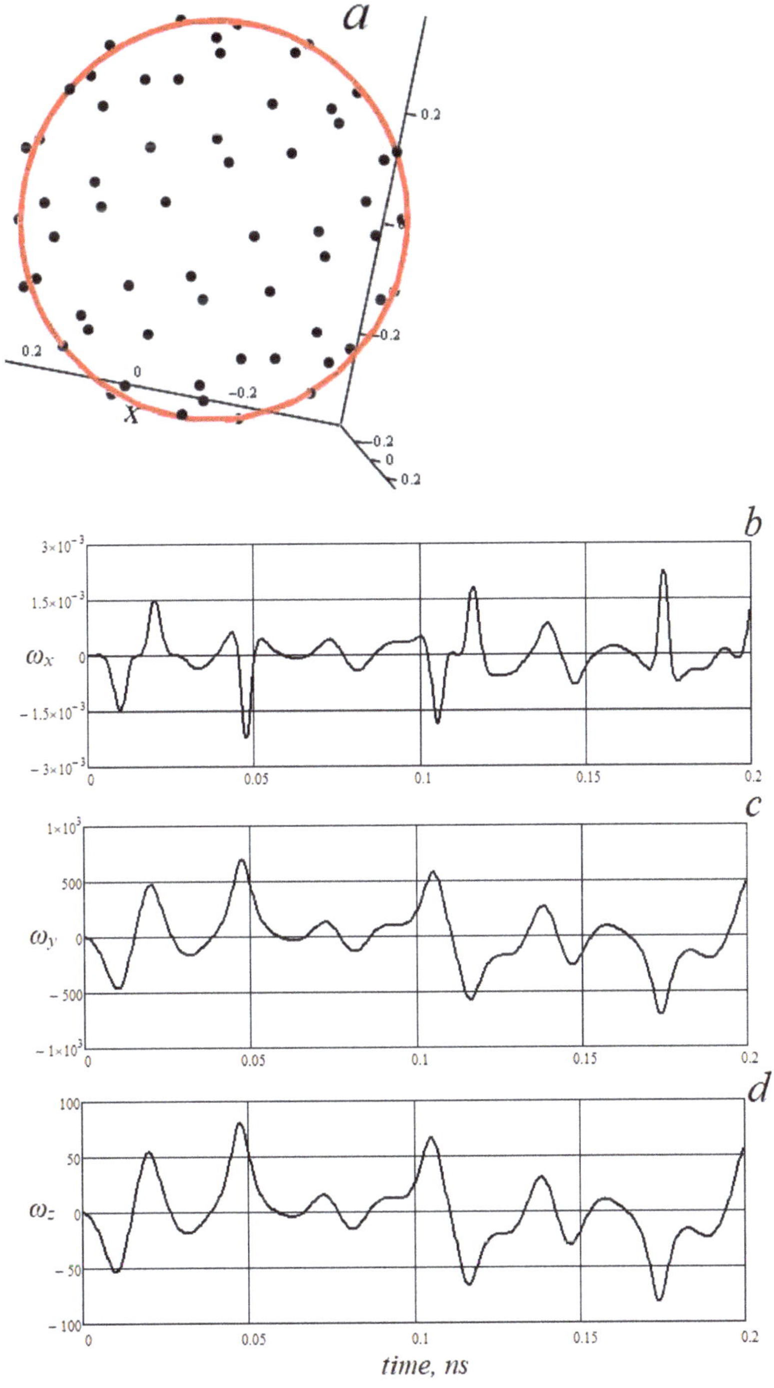

Figure 1. (**a**) Fullerene point trajectory; (**b**) x-projection of fullerene rotation frequency; (**c**) y-projection of fullerene rotation frequency; (**d**) z-projection of fullerene rotation frequency.

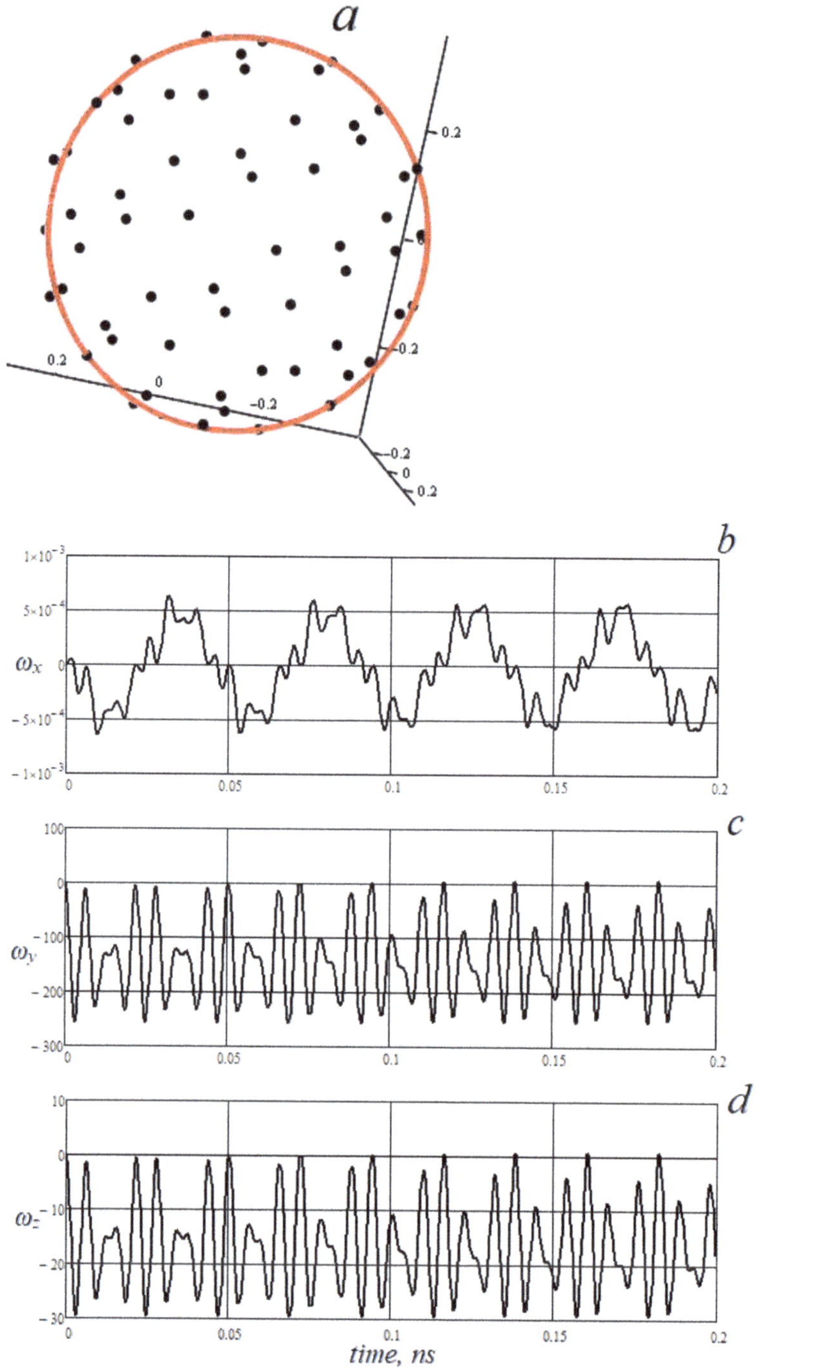

Figure 2. (**a**) Fullerene point trajectory; (**b**) x-projection of fullerene rotation frequency; (**c**) y-projection of fullerene rotation frequency; (**d**) z-projection of fullerene rotation frequency.

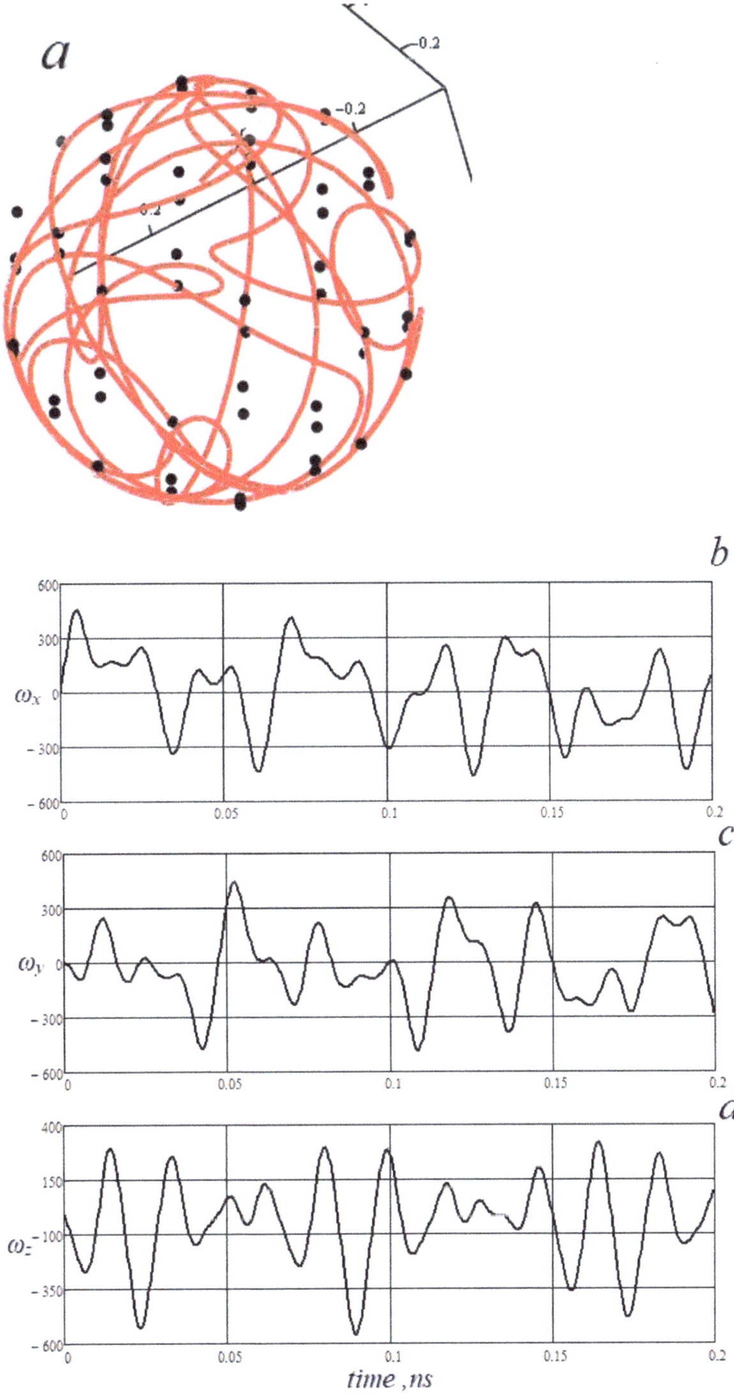

Figure 3. (**a**) Fullerene point trajectory; (**b**) x-projection of fullerene rotation frequency; (**c**) y-projection of fullerene rotation frequency; (**d**) z-projection of fullerene rotation frequency.

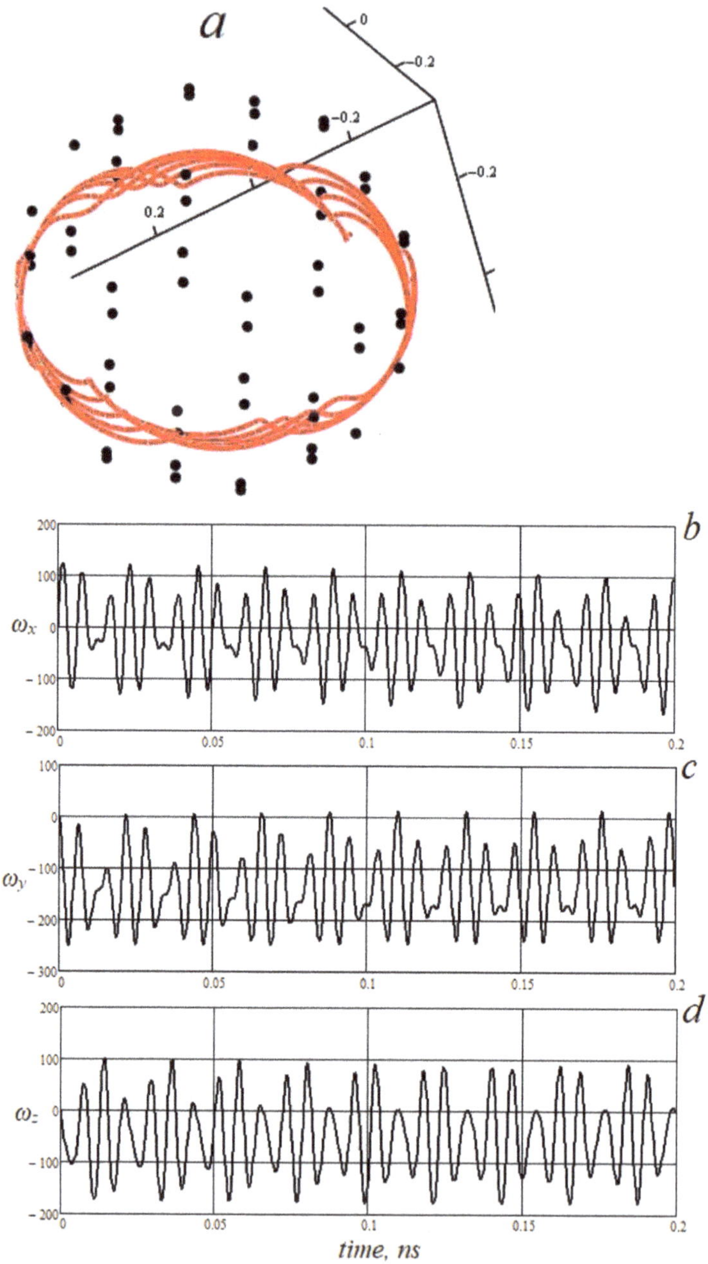

Figure 4. (**a**) Fullerene point trajectory; (**b**) x-projection of fullerene rotation frequency; (**c**) y-projection of fullerene rotation frequency; (**d**) z-projection of fullerene rotation frequency.

Figure 4 refer to the quasi-plane fullerene rotation in a rotating field f = 1000 GHz. The red line shows the trajectory of a representative carbon atom located at the initial time in the plane of vectors $\boldsymbol{B}$ and $\boldsymbol{\mu}$. By integrating the angular velocity over the section of unidirectional rotation, one can determine the amplitude of the resulting quasi-periodic oscillations. It turned out to be close to 2π. Therefore, we called this mode of motion

2π-rotations. It is interesting to note that somersaults that occur during an unstable rotation of a body around an intermediate axis (Louis Poinsot instability) are π rotations of the body around an axis with a minimum value of the moment of inertia. Rice. 2. demonstrates the nature of the movement of a magneto-susceptible fullerene at a magnetic field frequency f = 1000 GHz. As can be seen from the distributions of the components of the instantaneous angular velocity vector at moments close to the initial one, the fullerene motion is quasi-periodic, then it becomes periodic with time. Due to the initial direction of the intrinsic magnetic moment vector, chosen randomly, the plane of motion of the carbon atoms that make up the fullerene has slightly deviated from the x_z plane. This can be seen from the projection ω_z, which is synchronous with the projection ω_y and is slightly more than ten percent of it. In addition, both of these projections have the same sign. This means that the corresponding angular coordinates will change monotonically. Thus, the movement, in this case, will be a rotation (albeit uneven) around an axis that has a constant direction in space. Figure 3 shows the oscillations of a magnetic pendulum in a rotating magnetic field corresponding to the frequency f = 100 GHz. In this case, the components of the vector of the angular velocity of the magnetic fullerene are not synchronized and are equivalent in magnitude. As can be seen from Figure 4, at a frequency f = 1000 GHz in a rotating magnetic field, motions of a molecular object are obtained with the trajectory of an atomic particle on the surface of fullerene close to a certain plane determined by the initial data of magnetic quantities. In this case, the instantaneous angular velocity vector has significant deviations from the normal to the specified surface. However, as can be seen from Figure 4. there is a thickening of the trajectories, and we named this regime quasi-plane rotation.

4. Discussion

The desire to determine the position of a body in space by the angles of its rotation relative to some absolute basis inevitably leads to problems associated with restrictions on nutation angles in Euler's approach or to a complex algebra of hypercomplex numbers in Hamilton's approach. In this paper, we propose to determine the position of a non-deformable body rotating in space by a set of coordinates of its representative points (the trajectories of these points). In this work, we combined the tractor approach with a high-precision scheme for calculating evolutionary equations and obtained a calculation error in the test problem of a spherical magnetic pendulum close to machine accuracy. Based on the computational procedure developed by the authors, the possibility of spinning C_{60} fullerene intercalated by an iron atom by an alternating magnetic field was analyzed. It is found that angular vibrations of fullerene are observed at the field frequency f = 100 GHz. At a frequency above the specified value, the oscillations turn into directed rotations.

The proposed approach can be used to analyze the rotational dynamics of any body. In particular, it has wide applications in molecular dynamics since it can be generalized to the case of deformable molecular structures.

Author Contributions: Data curation, A.L.-F. and A.B.; Formal analysis, V.B.; Funding acquisition, M.B.; Investigation, M.B.; Methodology, A.B.; Project administration, M.B.; Resources, V.B.; Software, D.M.; Supervision and A.L.-F.; Validation, M.B.; Visualization, A.B.; Writing—original draft, D.M.; Writing—review & editing, D.M. All authors have read and agreed to the published version of the manuscript.

Funding: This research was funded by Russian Science Foundation under grant, No. 19 71 10049 and The APC was funded by Russian Science Foundation under grant, No. 19-71-10049.

Institutional Review Board Statement: Not applicable.

Informed Consent Statement: Not applicable.

Data Availability Statement: All data were obtained on the basis of a mathematical model described in detail in the paper. The code that implements the work, as well as all the initial and output data, the authors will submit upon request by email.

Acknowledgments: This research was supported by Russian Science Foundation under grant, No. 19-71-10049.

Conflicts of Interest: The authors declare no conflict of interest.

References

1. Fowler, P.W.; Manolopoulos, D.E. *An Atlas of Fullerenes*; Clarendon: Oxford, UK, 1995.
2. Guan, J.; Jin, Z.; Zhu, Z.; Chuang, C.; Jin, B.Y.; Tománek, D. Local curvature and stability of two-dimensional systems. *Phys. Rev. B* **2014**, *90*, 245403. [CrossRef]
3. Harris, P.J.F. Fullerene Polymers: A Brief Review. *C—J. Carbon Res.* **2020**, *6*, 71. [CrossRef]
4. Thakral, S.; Mehta, R.M. Fullerenes: Introduction and Overview of Their Biological Properties. *Indian J. Pharm. Sci.* **2006**, *68*, 13. [CrossRef]
5. Wang, L.; Wang, Y.; Hao, J.; Dong, S. Magnetic fullerene-DNA/hyaluronic acid nanovehicles with magnetism/reduction dual-responsive triggered release. *Biomacromolecules* **2017**, *18*, 1029–1038. [CrossRef]
6. Mallik, S.; Mattauch, S.; Dalai, M.K.; Brückel, T.; Bedanta, S. Effect of magnetic fullerene on magnetization reversal created at the Fe/C_{60} interface. *Sci. Rep.* **2018**, *8*, 5515. [CrossRef] [PubMed]
7. Lee, G.H.; Huh, S.H.; Jeong, J.W.; Ri, H.C. Excellent magnetic properties of fullerene encapsulated ferromagnetic nanoclusters. *J. Magn. Magn. Mater.* **2002**, *246*, 404–411. [CrossRef]
8. Takenobu, T.; Chi, D.H.; Margadonna, S.; Prassides, K.; Kubozono, Y.; Fitch, A.N.; Kato, K.; Iwasa, Y. Synthesis, structure, and magnetic properties of the fullerene-based ferromagnets Eu_3C_{70} and Eu_9C_{70}. *J. Am. Chem. Soc.* **2003**, *125*, 7, 1897–1904. [CrossRef]
9. Giacalone, F.; Martin, N. Fullerene polymers: Synthesis and properties. *Chem. Rev.* **2006**, *106*, 12, 5136–5190. [CrossRef]
10. Bhakta, P.; Barthunia, B. Fullerene and its applications: A review. *J. Indian Acad. Oral. Med. Radiol.* **2020**, *32*, 159–163. [CrossRef]
11. Bubenchikov, A.M.; Bubenchikov, M.A.; Mamontov, D.V.; Lun-Fu, A.V. Md-Simulation of Fullerene Rotations in Molecular Crystal Fullerite. *Crystals* **2019**, *9*, 496. [CrossRef]
12. Bubenchikov, A.M.; Bubenchikov, M.A.; Lun-Fu, A.V.; Ovchinnikov, V.A. Gyroscopic effects in fullerite crystal upon deformation. *Eur. Phys. J. Plus* **2021**, *136*, 388. [CrossRef]
13. Bubenchikov, M.A.; Bubenchikov, A.M.; Lun-Fu, A.V.; Ovchinnikov, V.A. Dynamics of Fullerenes in the Molecular Crystal of Fullerite. *Phys. Status Solidi A* **2021**, *218*, 2000174. [CrossRef]
14. Bubenchikov, M.; Ovchinnikov, V.; Mamontov, D. Numerical study of precession characteristics of rotating fullerene-C_{60} in intermolecular interaction. *AIP Conf. Proc.* **2020**, *2212*, 020013. [CrossRef]
15. Bubenchikov, M.A.; Lun-Fu, A.V.; Ovchinnikov, V.A. Precession of fullerite rotating node at simplest deformations of crystal fragment. *AIP Conf. Proc.* **2021**, *2337*, 020005. [CrossRef]
16. Bubenchikov, A.M.; Bubenchikov, M.A.; Ovchinnikov, V.A. Gyroscopic effect in intermolecular collision with the C_{60} fullerene molecule. *J. Phys. Conf. Ser. IOP Publ.* **2020**, *1537*, 012007. [CrossRef]
17. Sepahvand, S.; Farhadi, S. Fullerene-modified magnetic silver phosphate ($Ag_3PO_4/Fe_3O_4/C_{60}$) nanocomposites: Hydrothermal synthesis, characterization and study of photocatalytic, catalytic and antibacterial activities. *RSC Adv.* **2018**, *8*, 10124–10140. [CrossRef]
18. Bubenchikov, A.M.; Bubenchikov, M.A.; Mamontov, D.V.; Chelnokova, A.S.; Chumakova, S.P. Movement of fullerenes and their dimers inside carbon nanotubes. *Fuller Nanotub. Carbon Nanostructures* **2021**, *29*, 803–809. [CrossRef]

crystals MDPI

Article

Numerical Study of the Passage of Natural Gas Components through C_{60} Fullerite in the Low-Temperature Phase

Vladislav I. Borodin [1], Alexey M. Bubenchikov [2], Mikhail A. Bubenchikov [2], Vyacheslav A. Ovchinnikov [3,*] and Anna S. Chelnokova [2]

1 LLC "Gazprom Transgaz Tomsk", 9 Frunze St., Tomsk 634029, Russia
2 Department of Theoretical Mechanics, National Research Tomsk State University, 36 Lenin Ave., Tomsk 634050, Russia
3 Department of Physical and Computational Mechanics, National Research Tomsk State University, 36 Lenin Ave., Tomsk 634050, Russia
* Correspondence: empiric@mail.ru

Abstract: The movement of natural gas components through a crystalline fragment of C_{60} fullerite in a simple cubic phase is numerically investigated. The methods of classical molecular physics, the Monte Carlo and Euler approaches, and a step-by-step numerical method of a high order of accuracy are used to study the effects arising from the passage of the particles through the C_{60} fullerite. The influence of the characteristics of incoming particles on the permeability of fullerite was analyzed using the results of the performed calculations. The application of the fast integral approach based on the smoothed spherical potential is discussed and compared with the Monte Carlo method.

Keywords: rotating fullerene; natural gas; permeability; fullerite crystal; low temperature phase; Monte Carlo method

Citation: Borodin, V.I.; Bubenchikov, A.M.; Bubenchikov, M.A.; Ovchinnikov, V.A.; Chelnokova, A.S. Numerical Study of the Passage of Natural Gas Components through C_{60} Fullerite in the Low-Temperature Phase. *Crystals* **2022**, *12*, 1653. https://doi.org/10.3390/cryst12111653

Academic Editor: Philip Lightfoot

Received: 30 October 2022
Accepted: 15 November 2022
Published: 17 November 2022

1. Introduction

Since their discovery in 1985, fullerenes have been of great interest to researchers [1]. This is due to a wide range of unique properties (increased strength, hardness, high tribological properties, high fire and chemical resistance, durability, thermal stability, etc.) of fullerene-containing materials [2–11]. These properties make fullerenes interesting as a building block for creating structural materials with new properties [12–14]. Fullerenes are closed molecules consisting of 20 or more atoms located at the vertices of a polyhedron [15]. The most common fullerene is C_{60}, a truncated icosahedron built from 20 hexagons and 12 pentagons. At room temperature, these fullerenes form into a fullerite crystal, with a face-centered cubic (fcc) structure with a lattice parameter of 14.154 Å [16]. At temperatures below 260 K, the fullerite crystal transforms into a simple cubic (sc) phase with a lattice parameter of 14.111 Å [16]. In the range from 252–257 K, the fullerite structure is in a transitional phase, combining the fcc and sc phases [16]. In the fcc and sc phases, fullerenes perform rotational motions with frequency in the gigahertz range [17,18]. In the fcc phase, the centers of fullerene molecules are at a distance of 1.002 nm from each other [16]. This significantly complicates the movement of particles through the existing voids between fullerenes compared to the sc phase. Thus, a change in temperature makes it possible to radically change the permeability of C_{60} fullerite for incoming particles.

Natural gas mainly (70–98%) consists of methane (CH_4) and contains other heavy hydrocarbon gases (ethane, propane, butane, etc.) to a lesser extent. In natural gas, there is a fairly high content of rare inert monatomic gases such as helium, xenon, and others (many orders of magnitude higher than Earth's atmosphere). Helium (He) is the lightest noble gas and has a number of unique properties, such as chemical and radiological inertness, low density, high heat capacity and thermal conductivity, the lowest boiling point, etc. Xenon (Xe) is the heaviest noble gas and is used in spacecraft ion thrusters and as an effective and safe general anesthetic agent, but its high cost limits its use. In this regard, the possibility of separating these components from natural gas is of interest.

In this work, we numerically simulate the motion of incoming particles (components of natural gas) in a crystalline fragment of C_{60} fullerite in the sc phase. The numerical study is carried out using classical molecular dynamics methods [19–24] and the Monte Carlo method [25–27]. The Monte Carlo approach is a way to solve a problem numerically based on random statistical trials. This method is based on using a probabilistic model of the interaction of nano-objects and allows us to find effects that are difficult to identify using measurements. The additive and integral approaches are used to model incoming particles' interaction and fullerenes [19,23]. A comparative study of the influence of the characteristics of incident particles (helium, methane, ethane, xenon) and the parameters of the pair interaction potential on the permeability of fullerite is carried out.

2. Physical Statement of the Problem

The study of xenon atom motion in C_{60} fullerite is of the greatest interest since it has the lowest thermal velocity, which limits the probability of xenon atoms passing through fullerite compared to other natural gas components. The C_{60} fullerenes, depending on the temperature, can form sc and fcc structures. The fcc structures have the highest packing density of identical non-overlapping spheres in three-dimensional space. Therefore, the space unoccupied by C_{60} fullerenes will be minimal in the fcc lattice. The distance between the centers of the nearest fullerenes in the fcc fullerite (which is formed at temperatures above 260 K) is 1.002 nm, and this distance in the sc lattice (below 260 K) is 1.4111 nm. At the same time, the fcc lattice does not have straight tunnels, in contrast to the sc lattice shown in Figure 1. Taking into account the radius of C_{60} fullerene (a = 354 pm), as well as the van der Waals radii of the xenon atom (216 pm) and the carbon atom (170 pm), the passage of xenon atoms through the fcc structure of C_{60} fullerite has a very low probability, in contrast to fullerite in the sc phase.

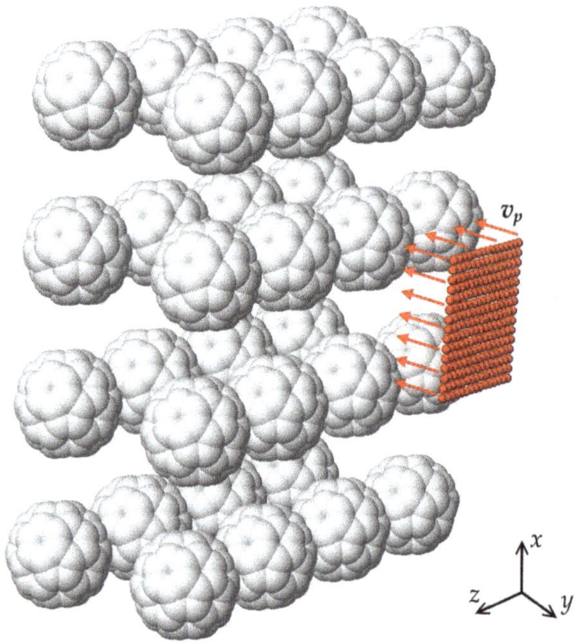

Figure 1. Scheme of the interaction of incoming particles (red color) with a group of rotating C_{60} molecules (grey color).

We consider a crystalline fragment of C_{60} fullerite, consisting of 32 fullerene molecules, as shown in Figure 1. The C_{60} molecules are located at the nodes of an sc lattice and interact with the incoming components of natural gas (helium, methane, ethane, xenon) (see Figure 1).

Each fullerene molecule is initially considered as an icosahedral structure consisting of sixty carbon atoms (additive approach) or as a particle with a spherical interaction potential (integral approach) [23]. The rotational motion of a fullerene is determined by the nature of the interaction of its carbon atoms with the surrounding fullerenes and is described using the Euler approach presented in papers [19,21,23]. The centers of mass of the nearest rotating C_{60} molecules are at a distance of a_0 = 1.4111 nm [16], corresponding to an sc phase of fullerite. The conditional beam of independent particles moves in the direction of the fullerite fragment. The interaction of carbon atoms of individual fullerenes with natural gas components is described based on the additive potential 12-6 repulsion attraction or the interaction potential in the surface-integral form.

3. Mathematical Statement of the Problem

Let us introduce some absolute (fixed) system of Cartesian coordinates (xyz), whose origin will be placed at the center of a series of fullerenes closest to the incoming particles. Cartesian coordinate systems ($\xi_i\eta_i\zeta_i$, $i = \overline{1,32}$) are rigidly connected with rotating fullerenes and their centers of mass.

If fullerenes are considered as structures consisting of 60 carbon atoms, the interaction of a particular carbon atom with an incoming particle or atoms of neighboring fullerenes is determined using the potential: 12-6 repulsion–attraction (additive approach)

$$U_{LJ}(\rho_{pik}) = 4\varepsilon\left[\left(\frac{\sigma}{\rho_{pik}}\right)^{12} - \left(\frac{\sigma}{\rho_{pik}}\right)^{6}\right], \tag{1}$$

where ρ_{pik} is the distance between the pth incoming particle and the kth atom of the ith fullerene, $i = \overline{1,S}$, $p = \overline{1,N}$, $k = \overline{1,K}$; S = 32 is the number of fullerenes of the crystalline fragment of fullerite; K = 60 is the number of carbon atoms in C_{60} fullerene; N = 225 is the number of incident particles of the conditional beam; ε, σ are the Lennard-Jones parameters [28].

If fullerenes are considered as spherical particles of radius a (integral approach), then the interaction potential between the ith fullerene and the pth incoming particle is determined by the following formula [19,23,29–31]

$$U_{sph}(\rho_{pi}) = \frac{4\varepsilon a\pi}{\rho_{pi}A_1}\left[\frac{\sigma^{12}}{5}\left((\rho_{pi}-a)^{-10} - (\rho_{pi}+a)^{-10}\right) - \frac{\sigma^6}{2}\left((\rho_{pi}-a)^{-4} - (\rho_{pi}+a)^{-4}\right)\right], \tag{2}$$

where $A_1 = A/K$, where $A = 4\pi a^2$ is the area of a sphere of radius a; ρ_{pi} is the distance between the center of the ith fullerene and the pth incoming particle. Potential (2) is based on the Lennard-Jones potential (1) uniformly distributed over the area of the sphere.

The rotational motion of the ith fullerene molecule around its center of mass is described using the Euler dynamic equations in the following form [21,23]

$$J_1\frac{dp_i}{dt} + (J_3 - J_2)q_i r_i = M_{\xi i}, \tag{3}$$

$$J_2\frac{dq_i}{dt} + (J_1 - J_3)r_i p_i = M_{\eta i}, \tag{4}$$

$$J_3\frac{dr_i}{dt} + (J_2 - J_1)p_i q_i = M_{\zeta i}. \tag{5}$$

here p_i, q_i, r_i are the components of the angular velocity relative to the principal axes; J_1, J_2, J_3 are the principal moments of inertia of the C_{60} molecule for its center of mass; $M_{\xi i}$, $M_{\eta i}$,

$M_{\zeta i}$ are the projections of the moments of external forces on the ξ-, η- and ζ-axes associated with the ith fullerene and determined according to the paper [23].

The components of the angular velocity of the ith fullerene molecule, written in terms of the Euler angles φ, ψ, θ, have the following form:

$$p_i = \dot{\psi}_i \sin\theta_i \sin\phi_i + \dot{\theta}_i \cos\phi_i, \quad (6)$$

$$q_i = \dot{\psi}_i \sin\theta_i \cos\phi_i - \dot{\theta}_i \sin\phi_i, \quad (7)$$

$$r_i = \dot{\psi}_i \cos\phi_i + \dot{\phi}_i, \quad (8)$$

where the dot above the function name indicate a derivative taken with respect to time.

The motion of the center of mass of the pth incoming particle when using the additive approach is determined by the law

$$m_p \frac{d\mathbf{v}_p}{dt} = \sum_{i=1}^{S} \sum_{k=1}^{K} gradU_{LJ}\left(\rho_{pik}\right), \; \mathbf{v}_p = \frac{d\boldsymbol{\rho}_p}{dt}, \quad (9)$$

and for integral approach

$$m_p \frac{d\mathbf{v}_p}{dt} = \sum_{i=1}^{S} gradU_{sph}(\rho_{pi}), \; \mathbf{v}_p = \frac{d\boldsymbol{\rho}_p}{dt}. \quad (10)$$

here m_p is the mass of the pth incoming particle; $\boldsymbol{\rho}_p$ is the position vector of the pth incoming particle. Note that since Equation (10) has one summation, the numerical solution of the problem will be much faster than Equation (9).

The initial conditions for solving the system of differential Equations (3)–(10) are given in the following form

$$t = 0: \; \psi_i = \psi_i^0, \; \theta_i = \theta_i^0, \; \varphi_i = \varphi_i^0; \; p_i = p_i^0, \; q_i = q_i^0, \; r_i = r_i^0, \quad (11)$$

$$\mathbf{v}_p = \mathbf{v}_p^0, \; \boldsymbol{\rho}_p = \boldsymbol{\rho}_p^0, \quad (12)$$

The system of differential Equations (3)–(10) with initial conditions (11), (12) is solved using a high-order step-by-step time integration scheme [32,33], which makes it possible to determine the spatial trajectories of the incoming particle and each atom of the central fullerene.

4. Results and Discussion

Consider a fragment of fullerite consisting of 4 × 4 × 2 = 32 fullerene molecules (see Figure 1), the initial orientation of which is set randomly (Monte Carlo method). We apply the Monte Carlo method only with an additive approach. The first row of 16 fullerenes is in position y = 0.0 nm; the second row is in position $y = -a_0 = -1.4111$ nm, where a_0 is the crystal lattice parameter. Incoming particles in the amount of 15 × 15 = 225 at the initial moment of time are in positions corresponding to y = 4.0 nm and have an initial velocity directed along the y-axis (Figure 1). In this paper, the initial velocity of the incoming particles is assumed to be equal to the value of the thermal velocity for the temperature T = 260 K, determined by the formula

$$v_p^0 = \sqrt{\frac{3k_B T}{m_p}}, \quad (13)$$

where k_B is the Boltzmann constant.

Figure 2 shows the mutual arrangement of fullerenes and a conditional beam of incident particles in the xy-plane. This number of particles makes it possible to estimate the fraction of transmitted particles with a step of ~0.44%. The conditional particle beam is located opposite four central fullerenes. This arrangement of incoming particles relative to fullerenes makes it possible to extend the obtained results to large scales. However, it should be taken into account that filling with incident particles with a similar step, for example, the space between the centers of 4 × 4 = 16 fullerenes, will require 43 × 43 particles, and not 9 × 225. This suggests that among the 225 selected particles, there is a small fraction of particles (~10%) that belong to neighboring regions. Therefore, it is necessary to apply appropriate coefficients when extending the results of calculations to large scales.

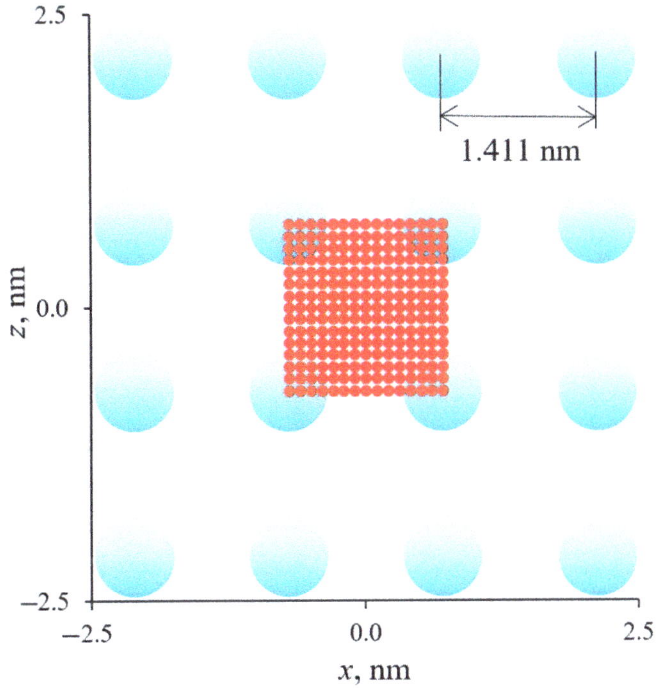

Figure 2. The initial position of the conditional beam of incoming particles (red color) relative to the considered C_{60} fullerite (blue color) in projection on the *xz*-plane.

Figure 3 shows a comparison of the additive ($U_\Sigma \times \varepsilon^{-1}$) and integral ($U_{sph} \times \varepsilon^{-1}$) dimensionless functions of the potential on the distance ρ_{pi} between the centers of the incoming particle and the center of the fullerene. The integral function of the potential U_{sph} is determined by Formula (2). The additive function of the Lennard-Jones potential U_Σ is determined by the following formula

$$U_\Sigma(\rho_{pi}) = \sum_{k=1}^{K} U_{LJ}\left(\rho_{pik}\right), \tag{14}$$

The potential integral function approximates the Lennard-Jones additive function well with an error of less than 1% at a distance between the particle and fullerene centers $\rho_{pi} > 0.705$ nm. The error increases significantly as the incoming particle and fullerene approach each other. The depth of the potential well differs by 4.5%.

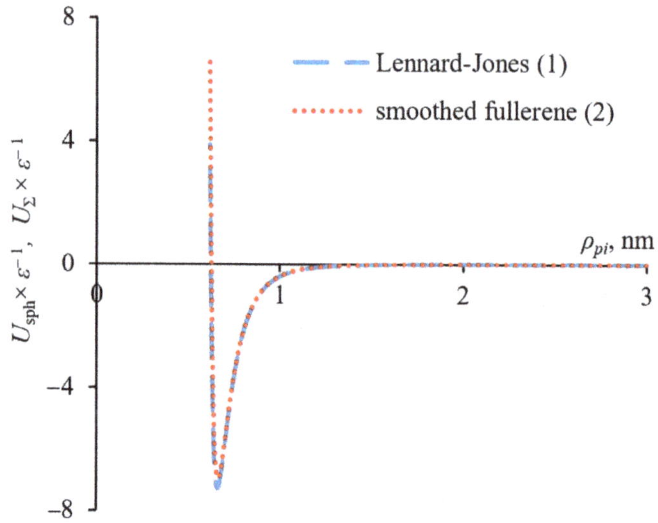

Figure 3. Dependences of relative potential energies for interaction between the C_{60} fullerene and single helium atom on the distance between the centers of mass.

4.1. Thermal Motion of Natural Gas Components

Let us consider when the initial velocities of incident particles (helium, methane, ethane, xenon) are equal to their thermal velocity. Using this assumption, we simulate the motion of particles near the surface of fullerite at a temperature of 260 K due to chaotic thermal motion ("thermal mode"). We will calculate up to the time $t_z = 0.02$ ns. Figure 4 shows that the sample size (200) is sufficient to determine the average value of the fraction of passing helium, methane, ethane, and xenon atoms.

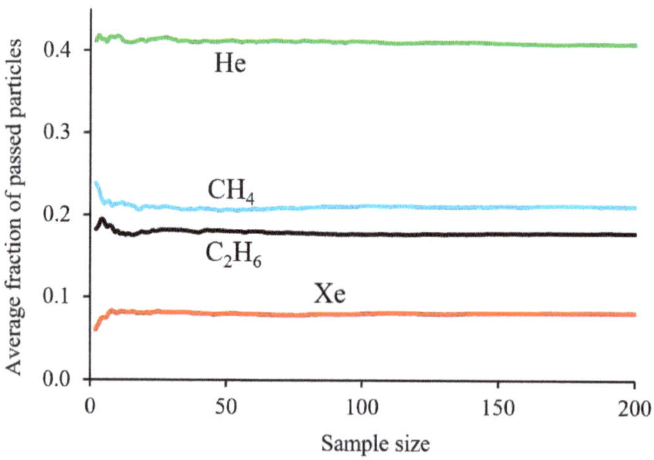

Figure 4. Dependence of the average value of the fraction of passed particles on the sample size.

Figure 5 shows the probability density function of the fraction of helium atoms that have passed through the fullerite fragment. The circles correspond to the calculations performed using the additive approach. The solid line shows the normal distribution obtained from the formula

$$\Phi(f) = \frac{1}{\sigma\sqrt{2\pi}} \exp\left(-\frac{1}{2}\left(\frac{f-\mu}{\sigma}\right)^2\right), \tag{15}$$

where the parameter μ is the mathematical expectation (mean), and σ is the standard deviation of the normal distribution.

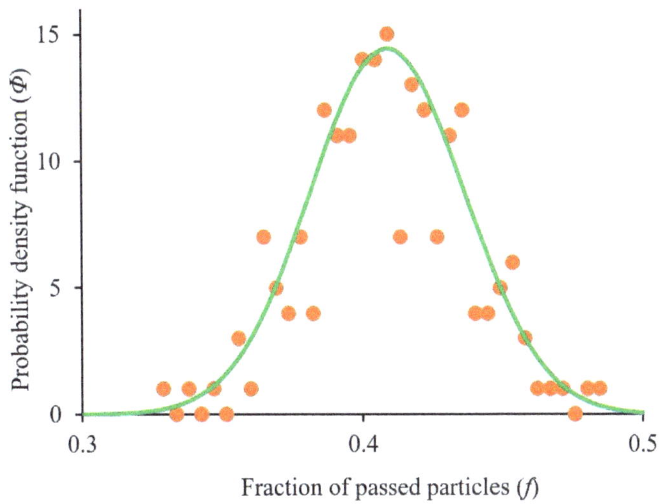

Figure 5. Probability density function for helium atoms. Orange color corresponds to calculations (additive approach). Green color corresponds to the normal distribution (15).

For the results presented in Figure 5, the mathematical expectation μ = 0.4089, and the standard deviation σ = 0.0276. As can be seen from Figure 5, the distribution of random variables obeys the normal distribution law.

The normal distributions (Figure 6) are similarly constructed for the other particles under consideration. Table 1 shows the interaction potential parameters ε and σ, obtained using the Lorentz-Berthelot mixing rules and used in Formulas (2) and (3). As can be seen from Figure 6, the larger the particle mass (Table 1), the less likely they are to pass through the fullerite crystalline fragment. Note that the time t_z = 0.02 ns is chosen so that all helium atoms leave the region occupied by fullerenes. However, according to (13), the initial velocity of other particles is significantly less than that of helium, so not all particles (methane, ethane, xenon) have time to leave the region occupied by fullerenes in the time interval $[0, t_z]$. This shows that helium has a significantly greater ability to overcome the crystalline fragment, unlike other particles. Note that xenon and helium atoms have the minimum and maximum thermal motion velocities, respectively. At the same time, the minimum and maximum fullerite permeability is observed for xenon and helium atoms, respectively. This suggests that the permeability of fullerite directly depends on the initial velocity of the particles (helium, methane, ethane, xenon) in thermal motion.

The dashed lines show the results of calculations obtained using the integral approach. As we can see, the integral approach to some extent, qualitatively reflects the relative pattern of permeability for different particles.

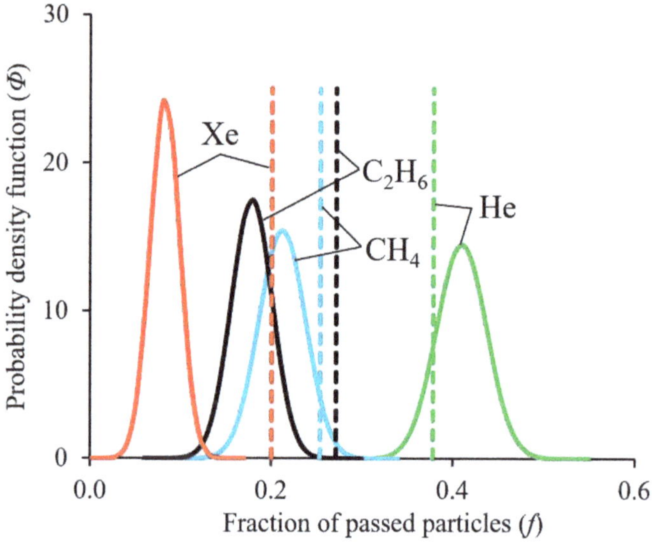

Figure 6. Probability density functions for particles moving with thermal velocity (13) corresponding to a temperature of 260 K.

Table 1. Lennard-Jones parameters for the mixture (particle-carbon) and particle mass.

Parameters	Mixture			
	He-C	**CH_4-C**	**C_2H_6-C**	**Xe-C**
ε/k_b	22.875	87.05	109.15	108.28
σ, nm	0.3102	0.3727	0.4024	0.3854
m_p, Da	4	16	30	131

4.2. Velocity Effect on the Permeability of Fullerite

In addition, we will analyze the permeability of fullerite under conditions of equal initial velocity, which is taken to be the thermal velocity of helium (see Figure 7). This assumption allows us to model the convective mode of movement of incoming particles near the fullerite surface. The average values of the fraction of passing atoms, determined using models using the additive and integral function of the potential, are presented in Table 1. As can be seen from Table 1, the best agreement between the methods was obtained for xenon (relative error δ = 2.7%) and the worst for methane and ethane (δ = 15.6–16.8%). At the same time, both methods show that helium and xenon give the largest number of passing particles, while methane and ethane give the smallest number. At the same time, it can be seen from Figure 7 and Table 2 that the probabilities of passage through fullerite, determined by the additive method, for helium and xenon turned out to be almost the same. However, calculations for helium showed large deviations from the average permeability compared to xenon. With the exception of xenon, a qualitative picture of the relative position of the distributions in Figures 6 and 7 has been preserved.

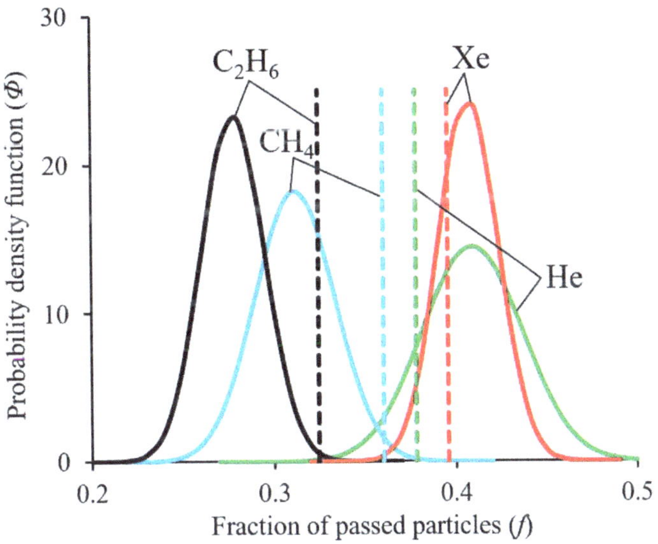

Figure 7. Probability density functions for particles moving with the same initial velocity v_p^0.

Table 2. Average values of the fraction of passed particles (f) calculated by the integral and additive methods.

Mean Values	Particles			
	He	CH_4	C_2H_6	Xe
f_{sph}	0.3778	0.36	0.3244	0.3956
f_Σ	0.4089	0.3115	0.2778	0.4067
δ	7.6%	15.6%	16.8%	2.7%

As can be seen from Table 1, the xenon atom, according to the Lennard-Jones parameters ε and σ, is between CH_4 and C_2H_6, i.e., not much different from them. At the same time, the fraction of passed xenon atoms is much larger than that of methane and ethane. This is due to the significantly larger mass of the xenon atom compared to methane and ethane.

The results of the calculations presented in Figures 6 and 7 show that with the convective movement of a natural gas substance near the fullerite, it is possible to significantly increase the probability of the passage of xenon atoms through the fullerite in the PS phase. The average probability of passage of xenon atoms increases from 8.22% (thermal mode) to 40.67% (convective mode). At the same time, changing the regime also increases the permeability of fullerite for methane (from 21.1% to 31.1%) and ethane (from 17.84% to 27.78%), but to a lesser extent. To assess the change in the concentration of a substance, one can analyze the selective release factor, which is defined as the ratio of the fractions of particles that have passed through fullerite. Based on the calculations obtained, the average selective xenon release factor with respect to methane (the main component of natural gas) in the thermal mode of particle motion near fullerite is 0.39. Therefore, fullerite will represent a barrier for xenon atoms and lead to a decrease in the concentration of xenon atoms. At the same time, this factor in the convective mode of movement of particles near fullerite is 1.31, which causes an increase in the concentration of xenon atoms after passing through the fullerite. This procedure must be carried out repeatedly to significantly increase the concentration of xenon atoms. As a result, this effect makes it possible to separate helium and xenon from natural gas using a fullerite crystal in the sc phase.

5. Conclusions

Classical molecular dynamics simulations were carried out to investigate the passage of natural gas components through the C_{60} fullerite in a simple cubic phase using the Monte Carlo method. The calculation results showed that natural gas components have different abilities to pass through fullerite. The analysis using the Monte Carlo method showed that helium atoms have the highest probability of passing through the fullerite barrier with a thermal velocity, and xenon atoms have the smallest one. However, at the same initial speed of movement, xenon atoms achieve a similar permeability of helium atoms. In addition, helium atoms have the smallest deviation from the average value of the fraction of the passed atoms. An increase in the initial velocity made it possible to significantly increase the permeability of fullerite for xenon. This effect can be used to increase the concentration of noble gases. It is shown that the method based on the integral potential makes it possible to quickly obtain a qualitatively correct overall picture of presented effects and the distribution of the average values of fullerite permeability for incoming particles.

Author Contributions: Conceptualization, A.M.B.; data curation, V.I.B.; formal analysis, M.A.B.; funding acquisition, M.A.B.; investigation, M.A.B. and V.A.O.; methodology, A.M.B.; project administration, M.A.B.; software, V.A.O. and A.S.C.; supervision, V.I.B.; visualization, V.A.O.; writing—original draft, V.A.O.; writing—review and editing, V.A.O. and A.M.B. All authors have read and agreed to the published version of the manuscript.

Funding: This research was funded by the Russian Federation President Grant, grant number MD-4273.2022.1.1.

Institutional Review Board Statement: Not applicable.

Informed Consent Statement: Not applicable.

Data Availability Statement: Not applicable.

Conflicts of Interest: The authors declare no conflict of interest.

References

1. Kroto, H.W.; Heath, J.R.; O'Brien, S.C.; Curl, R.F.; Smalley, R.E. C_{60}: Buckminsterfullerene. *Nature* **1985**, *318*, 162–163. [CrossRef]
2. Ruoff, R.S.; Ruoff, A.L. The Bulk Modulus of C_{60} Molecules and Crystals: A Molecular Mechanics Approach. *Appl. Phys. Lett.* **1991**, *59*, 1553–1555. [CrossRef]
3. Shen, H. The Compressive Mechanical Properties of C_n (n = 20, 60, 80, 180) and Endohedral M@C_{60} (M = Na, Al, Fe) Fullerene Molecules. *Mol. Phys.* **2007**, *105*, 2405–2409. [CrossRef]
4. Du, Z.; Jannatun, N.; Yu, D.; Ren, J.; Huang, W.; Lu, X. C_{60}-Decorated Nickel–Cobalt Phosphide as an Efficient and Robust Electrocatalyst for Hydrogen Evolution Reaction. *Nanoscale* **2018**, *10*, 23070–23079. [CrossRef] [PubMed]
5. Nisha, M.S.; Venthan, S.M.; Kumar, P.S.; Singh, D. Tribological Properties of Carbon Nanotube and Carbon Nanofiber Blended Polyvinylidene Fluoride Sheets Laminated on Steel Substrates. *Int. J. Chem. Eng.* **2022**, *2022*, 3408115. [CrossRef]
6. Ku, B.-C.; Han, Y.-C.; Lee, J.-E.; Lee, J.-K.; Park, S.-H.; Hwang, Y.-J. Tribological Effects of Fullerene (C_{60}) Nanoparticles Added in Mineral Lubricants According to Its Viscosity. *Int. J. Precis. Eng. Manuf.* **2010**, *11*, 607–611. [CrossRef]
7. Arie, A.A.; Chang, W.; Lee, J.K. Effect of Fullerene Coating on Silicon Thin Film Anodes for Lithium Rechargeable Batteries. *J. Solid State Electrochem.* **2010**, *14*, 51–56. [CrossRef]
8. Mackiewicz, N.; Bark, T.; Cao, B.; Delaire, J.A.; Riehl, D.; Ling, W.L.; Foillard, S.; Doris, E. Fullerene-Functionalized Carbon Nanotubes as Improved Optical Limiting Devices. *Carbon* **2011**, *49*, 3998–4003. [CrossRef]
9. Seo, K.-J.; Kim, D.-E. Molecular Dynamics Investigation on the Nano-Mechanical Behaviour of C_{60} Fullerene and Its Crystallized Structure. *Nanoscale* **2020**, *12*, 9849–9858. [CrossRef] [PubMed]
10. Smazna, D.; Rodrigues, J.; Shree, S.; Postica, V.; Neubüser, G.; Martins, A.F.; Ben Sedrine, N.; Jena, N.K.; Siebert, L.; Schütt, F.; et al. Buckminsterfullerene Hybridized Zinc Oxide Tetrapods: Defects and Charge Transfer Induced Optical and Electrical Response. *Nanoscale* **2018**, *10*, 10050–10062. [CrossRef]
11. Penkova, A.V.; Acquah, S.F.; Piotrovskiy, L.B.; Markelov, D.A.; Semisalova, A.S.; Kroto, H.W. Fullerene Derivatives as Nano-Additives in Polymer Composites. *Russ. Chem. Rev.* **2017**, *86*, 530–566. [CrossRef]
12. Horikawa, T.; Kinoshita, T.; Suito, K.; Onodera, A. Compressibility Measurement of C_{60} Using Synchrotron Radiation. *Solid State Commun.* **2000**, *114*, 121–125. [CrossRef]
13. Sundqvist, B. Fullerenes under High Pressures. *Adv. Phys.* **1999**, *48*, 1–134. [CrossRef]
14. Spitsina, N.G.; Motyakin, M.V.; Bashkin, I.V.; Meletov, K.P. C_{60} Fullerene and Its Molecular Complexes under Axial and Shear Deformation. *J. Phys. Condens. Matter* **2002**, *14*, 11089–11092. [CrossRef]

15. Terrones, H.; Terrones, M. Curved Nanostructured Materials. *New J. Phys.* **2003**, *5*, 126. [CrossRef]
16. Heiney, P.A.; Vaughan, G.B.M.; Fischer, J.E.; Coustel, N.; Cox, D.E.; Copley, J.R.D.; Neumann, D.A.; Kamitakahara, W.A.; Creegan, K.M.; Cox, D.M.; et al. Discontinuous Volume Change at the Orientational-Ordering Transition in Solid C_{60}. *Phys. Rev. B* **1992**, *45*, 4544–4547. [CrossRef]
17. Johnson, R.D.; Yannoni, C.S.; Dorn, H.C.; Salem, J.R.; Bethune, D.S. C_{60} Rotation in the Solid State: Dynamics of a Faceted Spherical Top. *Science* **1992**, *255*, 1235–1238. [CrossRef]
18. Johnson, R.D.; Yannoni, C.S.; de Vries, M.S. C_{60} Solid State Rotational Dynamics and Production and EPR Spectroscopy of Fullerenes Containing Metal Atoms. *Nanotechnology* **1992**, *3*, 164–166. [CrossRef]
19. Bubenchikov, A.M.; Bubenchikov, M.A.; Lun-Fu, A.V.; Ovchinnikov, V.A. Effect of a Charged Particle Propagating in C_{60} Fullerite. *Fuller. Nanotub. Carbon Nanostructures* **2021**, *29*, 442–445. [CrossRef]
20. Kang, J.W.; Hwang, H.J. Fullerene Shuttle Memory Device: Classical Molecular Dynamics Study. *J. Phys. Soc. Jpn.* **2004**, *73*, 1077–1081. [CrossRef]
21. Lun-Fu, A.V.; Bubenchikov, A.M.; Bubenchikov, M.A.; Ovchinnikov, V.A. Molecular Dynamics Study of Collective Behavior of Carbon Nanotori in Columnar Phase. *Crystals* **2021**, *11*, 1197. [CrossRef]
22. Sławianowski, J.J.; Kotowski, R.K. Classical Dynamics of Fullerenes. *Z. Für Angew. Math. Phys.* **2017**, *68*, 55. [CrossRef]
23. Lun-Fu, A.V.; Bubenchikov, A.M.; Bubenchikov, M.A.; Ovchinnikov, V.A. Computational Analysis of Strain-Induced Effects on the Dynamic Properties of C_{60} in Fullerite. *Crystals* **2022**, *12*, 260. [CrossRef]
24. González, M.A. Force Fields and Molecular Dynamics Simulations. *Éc. Thématique Société Fr. Neutron.* **2011**, *12*, 169–200. [CrossRef]
25. Kachour, Z.; Habchi, M.; Mesli, S.M.; Ziane, M.; Kotbi, M. On the Orientational Correlations in the Supercooled Chloride Lithium Aqueous Solution Using the Hybrid Reverse Monte Carlo Simulation. *Int. J. Mod. Phys. C* **2022**, *33*, 2250021. [CrossRef]
26. Liang, Y.; Muhammad, W.; Hart, G.R.; Nartowt, B.J.; Chen, Z.J.; Yu, J.B.; Roberts, K.B.; Duncan, J.S.; Deng, J. A General-Purpose Monte Carlo Particle Transport Code Based on Inverse Transform Sampling for Radiotherapy Dose Calculation. *Sci. Rep.* **2020**, *10*, 9808. [CrossRef]
27. Haghighat, A. *Monte Carlo Methods for Particle Transport*, 2nd ed.; CRC Press: Boca Raton, FL, USA, 2020; ISBN 978-0-429-19839-7.
28. Bubenchikov, M.A.; Bubenchikov, A.M.; Lun-Fu, A.V.; Ovchinnikov, V.A. Rotational Dynamics of Fullerenes in the Molecular Crystal of Fullerite. *Phys. Status Solidi A* **2021**, *218*, 2000174. [CrossRef]
29. Ruoff, R.S.; Hickman, A.P. Van Der Waals Binding to Fullerenes to a Graphite Plane. *J. Phys. Chem.* **1993**, *97*, 2494–2496. [CrossRef]
30. Baowan, D.; Hill, J.M. Mathematical Modeling of Interaction Energies between Nanoscale Objects: A Review of Nanotechnology Applications. *Adv. Mech. Eng.* **2016**, *8*, 168781401667702. [CrossRef]
31. Thamwattana, N.; Baowan, D.; Hill, J.M. Continuum Modelling for Interactions Between Fullerenes and Other Carbon Nanostructures. *J. Comput. Theor. Nanosci.* **2009**, *6*, 972–984. [CrossRef]
32. Ortega, J.M.; Poole, W.G. *An Introduction to Numerical Methods for Differential Equations*; Pitman: Mass, MA, USA, 1981; ISBN 978-0-273-01686-1.
33. Arora, G.; Joshi, V.; Garki, I.S. Developments in Runge–Kutta Method to Solve Ordinary Differential Equations. In *Recent Advances in Mathematics for Engineering*; Ram, M., Ed.; Series: Mathematical engineering, manufacturing, and management sciences; CRC Press: Boca Raton, FL, USA, 2020; pp. 193–202. ISBN 978-0-429-20030-4.

crystals

MDPI

Article

Synthesis and Characterization of Silver Nanoparticle-Polydimethylsiloxane (Ag-NP-PDMS) Stretchable Conductive Nanocomposites

Abdul Rauf Jamali [1], Jahanzeb Bhatti [1], Waseem Khan [1], Faheem Akther [2], Madiha Batool [3], Razia Batool [3] and Walid M. Daoush [4,5,*]

1 Department of Materials Engineering, NED University of Engineering and Technology, Karachi 75270, Pakistan
2 Department of Chemical Engineering, Quaid-e-Awam University of Engineering and Technology, Nawabshah 67480, Pakistan
3 Department of Chemistry, Government College University, Lahore 54000, Pakistan
4 Department of Production Technology, Faculty of Technology and Education, Helwan University, Cairo 11281, Egypt
5 Department of Chemistry, College of Science, Imam Mohammad ibn Saud Islamic University (IMSIU), Riyadh 11623, Saudi Arabia
* Correspondence: wmdaoush@imamu.edu.sa

Citation: Jamali, A.R.; Bhatti, J.; Khan, W.; Akther, F.; Batool, M.; Batool, R.; Daoush, W.M. Synthesis and Characterization of Silver Nanoparticle-Polydimethylsiloxane (Ag-NP-PDMS) Stretchable Conductive Nanocomposites. *Crystals* **2022**, *12*, 1098. https://doi.org/10.3390/cryst12081098

Academic Editor: Dmitri Donetski

Received: 28 June 2022
Accepted: 2 August 2022
Published: 5 August 2022

Abstract: A number of different research methodologies have been developed to increase the conductivity and mechanical properties of stretchable or flexible conductors. One of the promising techniques recommended for applying metallic nanoparticles (NPs) to PDMS (polydimethylsiloxane) substrate is to develop a thin-film that gives possible conductivity and good mechanical strain. This article discusses the preparation of silver nanoparticles using the chemical reduction method with silver nitrate as the precursor, and uses glucose as a reducing agent. In addition, polyvinyl pyrrolidone (PVP) is used to prevent the nanoparticles' oxidation and agglomeration once they have been synthesized successfully. Moreover, we utilize the power of diethylamine to accelerate the evolution of nanoparticles, and deionized water is used to prevent any possible contamination. The prepared Ag-NPs are then deposited on the solidified PDMS substrate through sintering. A multimeter is used to measure the electrical resistance. Ag-NPs are confirmed by UV-Vis at a 400-nm peak. Furthermore, we discuss the surface morphologies, particle sizes and thicknesses of the film and substrate when studied using different microscopy techniques. The prepared stretchable conductor is found to be suitable to use in biosensing and electronic devices.

Keywords: stretchable conductors; silver nanoparticles; polydimethylsiloxane; polyvinyl pyrrolidone (PVP); thin-film; coating; electrical resistance

1. Introduction

Stretchable electronics is a thrilling research area that combines numerous engineering fields such as materials, fabrication, electronics and mechanics. In stretchable electronics, the fabricated device maintains its electrical functionality under the application of axial, biaxial and/or repeated stretching loads [1]. This specific feature of the device could open a doorway for sensational and innovative applications of electronics such as artificial skin, strain sensors, artificial muscles or actuators, stretchable and flexible solar cells, and nanogenerators. Physical-sensing electronic devices that provide a sense of touch are receiving increased attention nowadays for futuristic applications [2–4].

Patterned structures of flexible, stretchable and electrically conductive materials on soft substrates could lead to the development of unique electronic devices with distinctive mechanical properties allowing them to bend, fold, stretch or fit the environmental conditions. Exploratory research on how we can improve the stretchability of circuits on elastomeric substrates has made outstanding progress. One of the promising advantages of large-area electronic devices is that they are thin and light enough to be placed easily on rooftops and walls. Besides their lightweight properties, the focus has now moved to their bending and rolling [5–7]. Large-area flexible sensors and actuator components, such as transistors and diodes, may be embedded in rubber sheets and joined with wavy metal wires by carefully monitoring the strains in thin-films. Their electrical circuits have high mechanical durability and show good electrical performance under stretching conditions as all the circuit components are stretchable [8–11]. To achieve this, a soft substrate is needed that includes many elastomers, i.e., silicon-based elastomers with versatile properties such as biocompatibility, flexibility, non-toxicity, hydrophobicity, and stretchability, etc. [12]. These samples are coated with thin-films of metals that can conduct electricity.

In this project, we used polydimethylsiloxane, commonly known as PDMS, as a soft elastomeric substrate because of its versatile properties such as stretchability, biocompatibility, bendability, etc. The key aim was to fabricate an elastic conductor that could conduct electricity under an applied strain. To do so, we synthesized silver nanoparticles that acted as a conducting thin-film on the PDMS substrate. A green nanotechnology route taking a chemical reduction method was developed to synthesize Ag-NPs using an environmentally friendly and low-cost method. The chemical reduction method has the advantage of effectively synthesizing nanoparticles both at the laboratory scale and when upscaled for mass production, as well. In our research, the prepared nanoparticles were then deposited on the PDMS substrate. PDMS was used to provide strong elastomeric properties against mechanical strain, in addition to its effect on the conductivity of the produced nanocomposites.

2. Materials and Methods

PDMS in powder form with 97% purity was used as a substrate, purchased from Dow Corning Corporation. Silver nitrate, used for the formation of Ag-NPs, was purchased from Merck (Pvt.) Ltd., Karachi, Pakistan. with a purity of 99.95%. Di-ethyl amine and glucose were purchased from the local market, both of commercial grade and 95% purity. PVP with an average of 35,000 mol. wt. was purchased from Sigma-Aldrich, Burlington, MA, USA.

2.1. *Synthesis of Ag-NPs*

Initially, PDMS substrates were fabricated using the solution processing method. The silicone elastomeric base and silicone elastomeric hardener were mixed with a ratio of 10:1 g. The prepared 2.5 g of PDMS solution was then poured into the mold (5 × 5 cm) to achieve 1 mm-thick PDMS [13–15]. The film thickness of the sample was measured by stereomicroscope (MOTIC DMW-143). The Ag-NPs (<100 nm) were then prepared chemically via reduction (a green nanotechnology approach) using the glucose-gluconic acid oxidation-reduction method in diethylamine, with silver nitrate as a source of silver ions. The prepared Ag-NPs were then deposited as a thin-film by inkjet printing and sintering at 100 °C for 30 min. A straightforward in-house technique was used to prepare the thin-film of the nanocomposites via the inkjet printing method. Finally, the samples were characterized through different techniques.

As mentioned, the Ag-NPs solution was prepared by chemical reduction (a green nanotechnology route). An aqueous solution of 100 mM silver nitrate and 50 mM glucose was mixed and stirred using a magnetic stirrer at room temperature for 20 min to obtain a homogeneous solution. An aqueous solution of 115 mM molarity diethylamine (DEA) was added, and the solution was mixed quickly and vigorously [16–18]. The color of the solution changed from yellow to brown and finally to black, a possible indication of Ag-NPs [19–21], as shown in Figure 1.

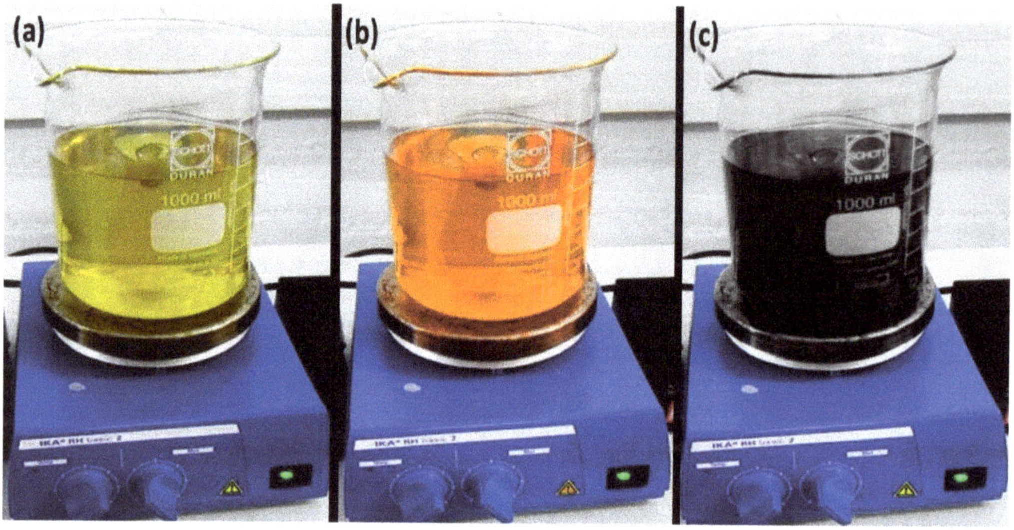

Figure 1. Synthesis process of Ag-NPs; (**a**) yellow color appeared during the mixing of silver nitrate and glucose with diethylamine (**b**) yellow color of the solution change to orange and (**c**) the orange color of the solution is turned to black indicating the formation of Ag-NPs

The recommended mechanism for this reaction is similar to the silver mirror technique, where the amine is dissolved in water to remove hydrogen ions, leaving hydroxyl ions in the solution. The hydrated amine ion further reacts with silver nitrate to form a complex of silver ions. The remaining hydroxyl ions oxidize the aldehyde groups of the glucose molecules to form gluconic acid, and an electron is released into the solution. This electron reduces the silver ions of the silver complex to get metallic nanoparticles of silver. Diethylamine (DEA) can be used as a catalyst. To prevent agglomeration of the particles, polyvinylpyrrolidone (PVP) is used as a capping agent, as shown in Figure 2. The maximum amount of 3 wt.% PVP is used in the solution, which acts as a stabilizer preventing the agglomeration of the synthesized nanoparticles. Small wt. percentage of PVP was used because the greater the added amount of PVP can makes the particles nonconductive. Nonetheless, when using PVP as the stabilizing agent, producing just small quantities of the synthesized particles is efficient to form the thin-films.

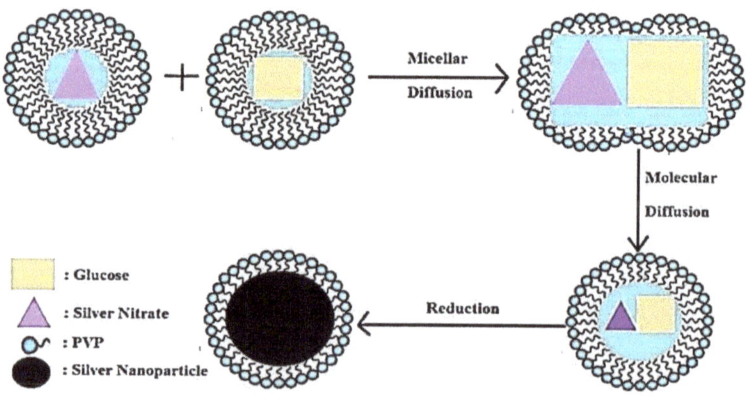

Figure 2. Mechanism of using PVP stabilizer for silver nanoparticles.

2.2. Preparation of Ag-NPs/Ethylene Glycol Ink

The prepared solution of silver nanoparticles was then used for ink preparation by centrifugation (at a 3000-rpm speed for 10 min at 220 volts), using ethylene glycol as a dispersing agent, as shown in Figure 3.

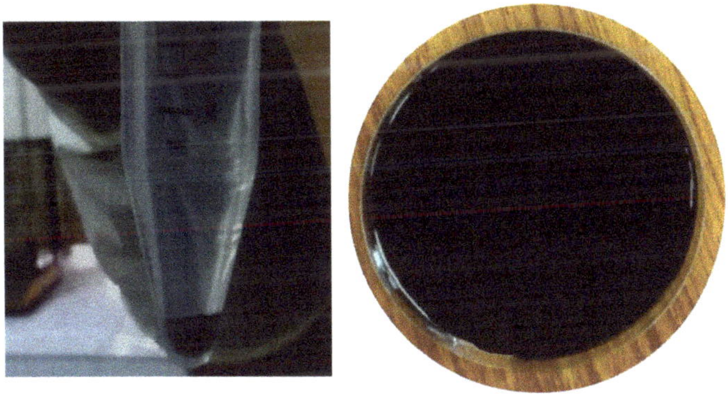

Figure 3. Ag-NPs after centrifugations (**left**). Silver nanoparticle ink dispersed in ethylene glycol (**right**).

The silver nanoparticle ink was then placed in a vial and subjected to jet ultrasonication, to break the bonds between agglomerated nanoparticles and thus disperse the nanoparticles in the freshly prepared ink. The nanoparticle ink was stored in a vial that underwent sonication using an ultrasonic cleaning machine.

2.3. Fabrication of Ag-NP Thin-Film

The prepared Ag-NPs were then deposited as a thin-film using the inkjet printing technique. The coated film on the substrate (PDMS/Ag-NPS) was then consolidated in the oven for 1 h at 100 °C, and cooled in a dry oven for good adhesion. Figure 4 shows the uncoated and coated substrate samples produced after solidification.

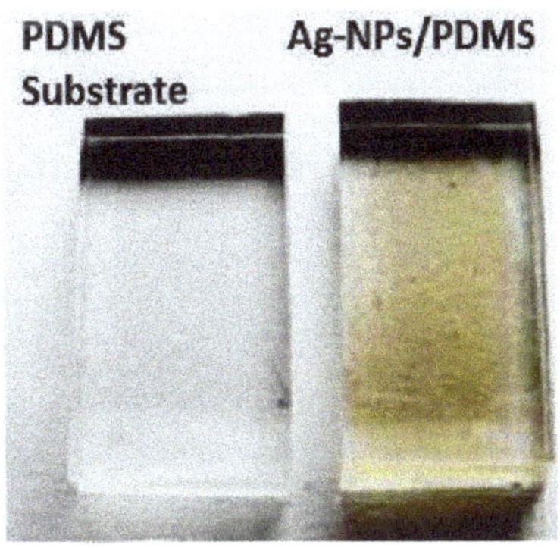

Figure 4. Thin-layer coats of Ag-NP ink on the PDMS substrate.

2.4. Characterization of the Ag-NPs and the Ag-NP-PDMS Thin-Film

The surface of the prepared samples was investigated using an Olympus Microscope model GX-51 (Center Valley, PA, USA), while scanning electron microscope (SEM) model Quanta 200 (Hillsboro, OR, USA) was used to determine the surface morphologies and particle sizes and shapes of the prepared Ag-NPs. A UV-Vis spectrophotometer model Spectrumlab 22PC (Shanghai. Lengguang Technology Co. Ltd., Shanghai, China) was used to determine the absorption spectrum of the prepared silver nanoparticles. The laboratory apparatus was designed to measure the strain versus the electrical resistance of the sintered thin-film samples when increasing the applied load, by assessing the electrical resistance of the sintered thin-films at different strain loads using a multimeter. The electrical measurements were recorded before and after applying each load [22,23].

3. Results and Discussion

3.1. Ag-NPs' Characterization

Investigations of the produced Ag-NPs were carried out using a field emission scanning electron microscope at a standard high voltage of 25–30 kV to study the particle shapes, sizes, and morphologies. Figure 5 shows SEM images with different magnifications of the prepared Ag-NPs. In Figure 5a, the particles are at the nanoscale and we can see some agglomerations. The particle size range is between 108 and 189.6 nm, and some particles are agglomerated in different areas. Undesirable agglomeration of particles results due to the conglomeration and pileup of the particles, resulting in the deterioration of the size range [24–26]. Figure 5b–d show SEM images with high magnifications, in which the phenomenon of particle agglomeration can be observed more clearly and effectively. The particle size of one particle was measured as 108.3–189.6 nm, but agglomeration appeared in various regions, resulting in vast increments in the size range. Particle pile-up was obvious in the region of the red box, resulting in a dramatic increase in the sizes of the particles [27].

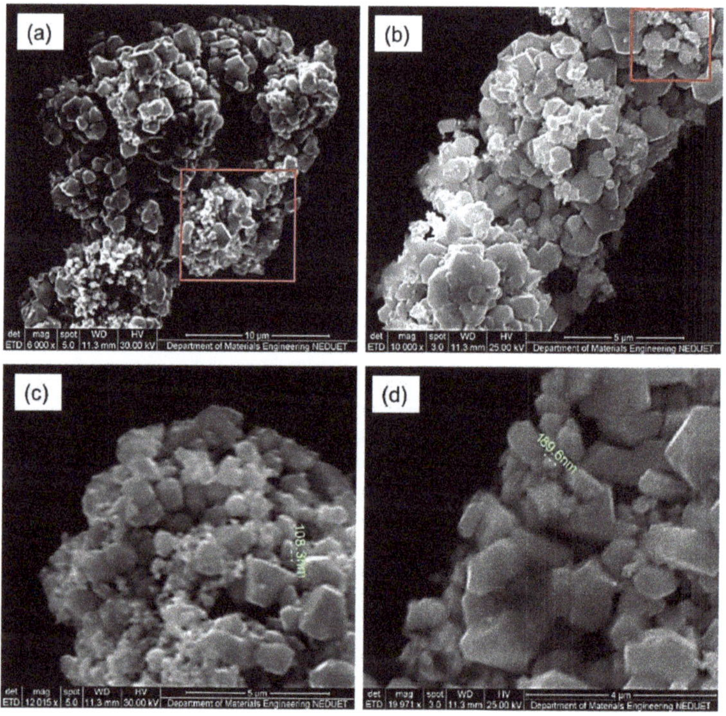

Figure 5. SEM images with different magnifications, where (**a**) 6000×, (**b**) 10,000×, (**c**) 12,015× and (**d**) 19,971× of the prepared Ag-NPs. The triangular shape of particles can be seen at different magnification.

3.2. UV Absorption Spectrum of the Ag-NPs

Figure 6 shows the spectrum of standard UV spectroscopy for the produced silver nanoparticles in the wavelength and absorbance range of UV light. The spectrum was scanned within the wavelength range from 380 to 420 nm. A high-intensity peak was detected at 400 nm, which confirmed the formation of Ag-NPs. It is believed that the UV-Vis absorption peak of Ag-NPs can be detected in the range of 390–470 nm, depending on the particle sizes, shapes, and distribution [28]. Hence, the absorption peak presented in Figure 6 demonstrates the presence of the Ag-NPs, as revealed when investigated by SEM (see Figure 5).

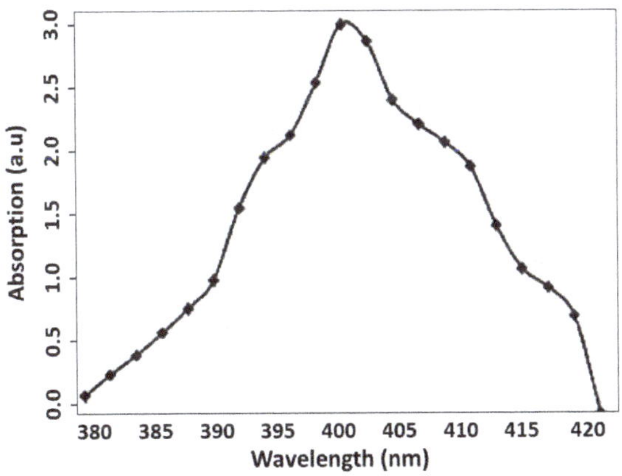

Figure 6. UV spectrum of the AgNPs dispersed in ethanol solution.

3.3. Surface Characterization of Ag-NP-PDMS Thin-Film

Figure 7a depicts the results of the stereomicrograph of the surface of the thin-film; the black spots show the silver particles and the grey background shows the substrate. The undesirable agglomeration phenomenon can be observed, confirmed by the presence of large black spots, such as in the lower left of the micrograph shown in Figure 7a. Figure 7b presents a high magnification of the surface. The zones in the red boxes in the images capture notable agglomeration [29]. Figure 7c reveals an important property of Ag-NPs, i.e., shining and brightening in the presence of light from the microscope.

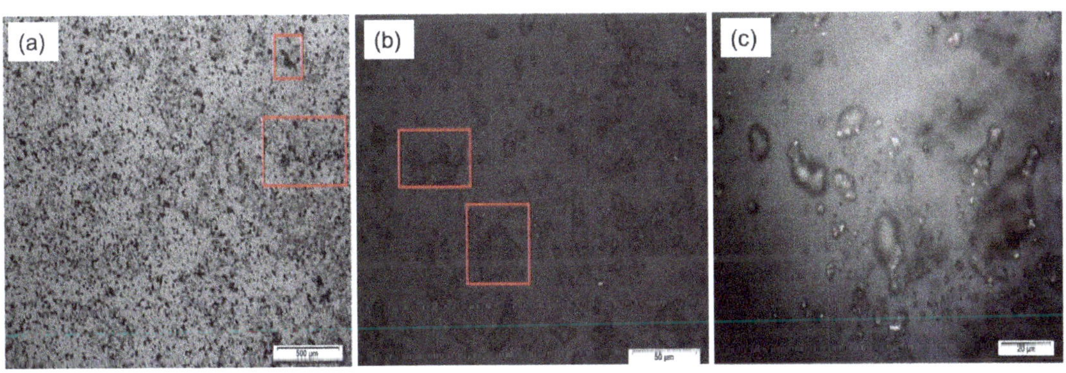

Figure 7. (**a–c**) Stereomicrographs with different magnifications of the fabricated thin-film.

The stereomicroscope technique is used to measure line spacing, as well as sample thickness. The results obtained from this test are shown below in Figure 8. There was great variation in the thicknesses of the samples when a mean of four readings was calculated. The variation ranged from 0.914 to 1.312 mm, with the thicknesses of the coated layers differing.

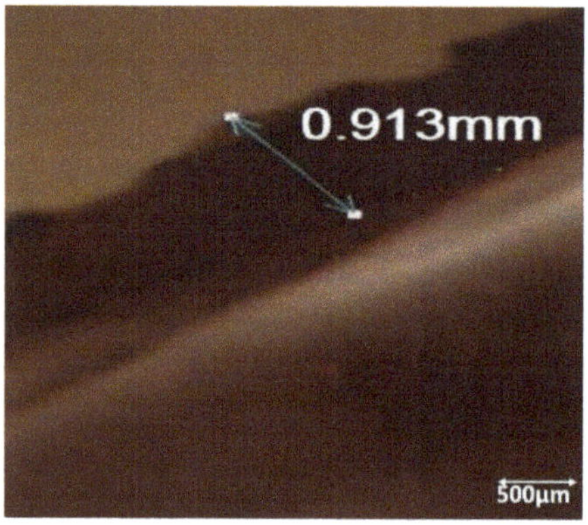

Figure 8. Stereomicrograph of the cross-sectional area of the deposited Ag-NP thin-film on the PDMS substrate.

Substrates with different roughness patterns were prepared and silver nanoparticles were synthesized. Then, in this step, the substrates were coated with silver nanoparticles, followed by a sintering process. Figure 9a shows a stereomicrograph of the prepared Ag-NP thin-film on the PDMS substrate sample after sintering. Following this, the sample was subjected to stretching using a homemade apparatus. Next, the electrical resistance of the prepared and stretched samples was measured after applying a suitable load. The sample surface after stretching was investigated using a stereomicroscope, as shown in Figure 9b. It was observed from the micrograph that the sintered Ag-NP thin-film on the PDMS substrate sample (Figure 9a) had a uniform surface morphology. However, the sintered Ag-NP thin-film on the PDMS substrate sample (Figure 9b), after stretching by applying a suitable load, had a deformed surface morphology with some areas marred by grooves, potentially due to the loss of some coated materials from the surface of the substrate during stretching [30].

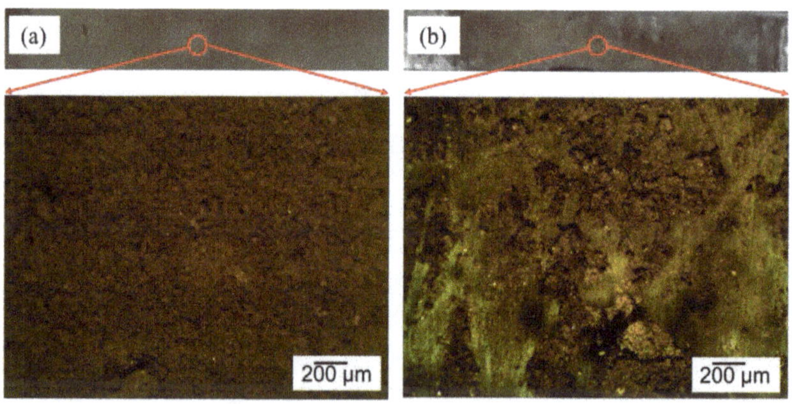

Figure 9. Stereophotographs of the sintered thin-film of Ag-NPs supported on the PDMS substrate, (**a**) before and (**b**) after stretching.

3.4. Electrical Conductivity of the Sintered Ag-NP-PDMS Thin-Film

After successful sintering of silver nanoparticles on the surface of the PDMS substrate, the coated samples were subjected to a measurement of their electrical conductivity under an applied load. Figure 10 shows the electrical resistance values of the thin-film under an applied load with respect to changes in the length of the thin-film. In both unstretched and stretched sample conditions, the electrical resistance of the sample in the horizontal directions was found to be the best, with the lowest electrical resistance [31]. For instance, when the unstretched sample was 2.4 cm in length, the electrical resistance was found to be 1.6 ohms. The sample was stretched slowly, and increases in electrical resistance were observed at higher values, with the highest electrical resistance value of 3.15 ohms at the length of 3.15 cm. In comparison, when we consider the electrical resistance measurements in the vertical and radial directions, the resistances of the unstretched samples were 2.5 and 3.8 ohms at sample lengths of 3.5 and 2.5 cm, respectively. When stretching the sample vertically, the electrical resistance value was observed to be 65 ohms for a maximum stretch of 3.1 cm [32]. Meanwhile, when stretched radially, the maximum electrical resistance was observed to be 73 ohms [33].

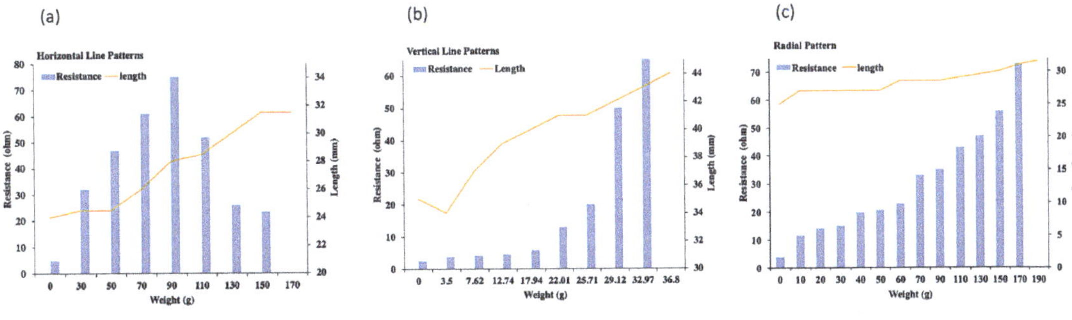

Figure 10. Electrical resistance vs. stretching length of (**a**) horizontal, (**b**) vertical and (**c**) radial patterns.

4. Conclusions

Silver nanoparticles were prepared chemically with a green reduction method using glucose-gluconic acid oxidation-reduction in diethylamine, with silver nitrate as a source of silver ions. The synthesized silver nanoparticles had an average particle size of 108.3–189.6 nm. Their presence was confirmed by UV-Vis spectrophotometry, indicating an intense peak at a wavelength of 400 nm. PDMS was used as a substrate in the liquid form, and it was solidified via sintering. The prepared silver nanoparticles were successfully deposited as a thin-film on the prepared PDMS substrates by inkjet printing, followed by sintering at 100 °C for 30 min. The electrical resistance of the obtained sintered samples was measured before and after stretching by applying a suitable load in different directions. The samples stretched horizontally showed the best electrical properties. Based on the obtained results, the prepared silver nanoparticle/PDMS stretchable conductor has effective electrical and mechanical properties to be used in bio-sensing and electronic devices. Future studies will be conducted to develop the technique and the properties of the obtained nanocomposites.

Author Contributions: Conceptualization, A.R.J., J.B., W.K., F.A., M.B., R.B. and W.M.D.; methodology, A.R.J., J.B., W.K., F.A., M.B., R.B. and W.M.D.; validation, A.R.J., J.B., W.K., F.A., M.B., R.B. and W.M.D.; formal analysis, A.R.J., J.B., W.K., F.A., M.B., R.B. and W.M.D.; investigation, A.R.J., J.B., W.K., F.A., M.B., R.B. and W.M.D.; resources, A.R.J., J.B., W.K., F.A., M.B., R.B. and W.M.D.; data curation, A.R.J., J.B., W.K., F.A., M.B., R.B. and W.M.D.; writing—original draft preparation A.R.J., J.B., W.K., F.A., M.B., R.B. and W.M.D.; writing—review and editing, A.R.J., J.B., W.K., F.A., M.B., R.B. and W.M.D.; visualization, A.R.J., J.B., W.K., F.A., M.B., R.B. and W.M.D.; supervision, W.M.D.; project administration, A.R.J. All authors have read and agreed to the published version of the manuscript.

Funding: This research received no external funding.

Institutional Review Board Statement: Not applicable.

Informed Consent Statement: Informed consent was obtained from all subjects involved in the study.

Data Availability Statement: Not applicable.

Conflicts of Interest: The authors declare no conflict of interest.

References

1. Abu-Khalaf, J.M.; Al-Ghussain, L.; Al-Halhouli, A.A. Fabrication of stretchable circuits on polydimethylsiloxane (PDMS) pre-stretched substrates by inkjet printing silver nanoparticles. *Materials* **2018**, *11*, 2377. [CrossRef] [PubMed]
2. Feng, P.; Ji, H.; Zhang, L.; Luo, X.; Leng, X.; He, P.; Zhao, W. Highly stretchable patternable conductive circuits and wearable strain sensors based on polydimethylsiloxane and silver nanoparticles. *Nanotechnology* **2019**, *30*, 185501. [CrossRef] [PubMed]
3. Soe, H.M.; Abd Manaf, A.; Matsuda, A.; Jaafar, M. Development and fabrication of highly flexible, stretchable, and sensitive strain sensor for long durability based on silver nanoparticles–polydimethylsiloxane composite. *J. Mater. Sci. Mater. Electron.* **2020**, *31*, 11897–11910. [CrossRef]
4. Min, S.H.; Lee, G.Y.; Ahn, S.H. Direct printing of highly sensitive, stretchable, and durable strain sensor based on silver nanoparticles/multi-walled carbon nanotubes composites. *Composites Part B Eng.* **2019**, *161*, 395–401. [CrossRef]
5. Guo, Y.; Yu, J.; Li, C.; Li, Z.; Pan, J.; Liu, A.; Zhang, C. SERS substrate based on the flexible hybrid of polydimethylsiloxane and silver colloid decorated with silver nanoparticles. *Opt. Express* **2018**, *26*, 21784–21796. [CrossRef]
6. Al-Halhouli, A.A.; Al-Ghussain, L.; El Bouri, S.; Liu, H.; Zheng, D. Fabrication and evaluation of a novel non-invasive stretchable and wearable respiratory rate sensor based on silver nanoparticles using inkjet printing technology. *Polymers* **2019**, *11*, 1518. [CrossRef]
7. Zhang, S.; Zhang, H.; Yao, G.; Liao, F.; Gao, M.; Huang, Z.; Lin, Y. Highly stretchable, sensitive, and flexible strain sensors based on silver nanoparticles/carbon nanotubes composites. *J. Alloys Compd.* **2015**, *652*, 48–54. [CrossRef]
8. Soe, H.M.; Abd Manaf, A.; Matsuda, A.; Jaafar, M. Performance of a silver nanoparticles-based polydimethylsiloxane composite strain sensor produced using different fabrication methods. *Sens. Actuators A Phys.* **2021**, *329*, 112793. [CrossRef]
9. Choi, Y.I.; Hwang, B.U.; Meeseepong, M.; Hanif, A.; Ramasundaram, S.; Trung, T.Q.; Lee, N.E. Stretchable and transparent nanofiber-networked electrodes based on nanocomposites of polyurethane/reduced graphene oxide/silver nanoparticles with high dispersion and fused junctions. *Nanoscale* **2019**, *11*, 3916–3924. [CrossRef]
10. Tavakoli, M.; Malakooti, M.H.; Paisana, H.; Ohm, Y.; Green Marques, D.; Alhais Lopes, P.; Majidi, C.E. GaIn-Assisted Room-Temperature Sintering of Silver Nanoparticles for Stretchable, Inkjet-Printed, Thin-Film Electronics. *Adv. Mater.* **2018**, *30*, 1801852. [CrossRef]
11. Shankar, A.; Salcedo, E.; Berndt, A.; Choi, D.; Ryu, J.E. Pulsed light sintering of silver nanoparticles for large deformation of printed stretchable electronics. *Adv. Compos. Hybrid Mater.* **2018**, *1*, 193–198. [CrossRef]
12. Chen, J.; Zheng, J.; Gao, Q.; Zhang, J.; Zhang, J.; Omisore, O.M.; Li, H. Polydimethylsiloxane (PDMS)-based flexible resistive strain sensors for wearable applications. *Appl. Sci.* **2018**, *8*, 345. [CrossRef]
13. Huang, Q.; Al-Milaji, K.N.; Zhao, H. Inkjet printing of silver nanowires for stretchable heaters. *ACS Appl. Nano Mater.* **2018**, *1*, 4528–4536. [CrossRef]
14. Htwe, Y.Z.N.; Hidayah, I.N.; Mariatti, M. Performance of inkjet-printed strain sensor based on graphene/silver nanoparticles hybrid conductive inks on polyvinyl alcohol substrate. *J. Mater. Sci. Mater. Electron.* **2020**, *31*, 15361–15371. [CrossRef]
15. Wang, T.; Wang, R.; Cheng, Y.; Sun, J. Quasi in situ polymerization to fabricate copper nanowire-based stretchable conductor and its applications. *ACS Appl. Mater. Interfaces* **2016**, *8*, 9297–9304. [CrossRef] [PubMed]
16. Abu-Khalaf, J.; Al-Ghussain, L.; Nadi, A.; Saraireh, R.; Rabayah, A.; Altarazi, S.; Al-Halhouli, A.A. Optimization of geometry parameters of inkjet-printed silver nanoparticle traces on pdms substrates using response surface methodology. *Materials* **2019**, *12*, 3329. [CrossRef]
17. Xin, Z.; Liu, Y.; Li, X.; Liu, S.; Fang, Y.; Deng, Y.; Li, L. Conductive grid patterns prepared by microcontact printing silver nanoparticles ink. *Mater. Res. Express* **2017**, *4*, 015021. [CrossRef]
18. Lee, J.Y.; Shin, D.; Park, J. Fabrication of silver nanowire-based stretchable electrodes using spray coating. *Thin Solid Film.* **2016**, *608*, 34–43. [CrossRef]
19. Duan, S.; Yang, K.; Wang, Z.; Chen, M.; Zhang, L.; Zhang, H.; Li, C. Fabrication of highly stretchable conductors based on 3D printed porous poly (dimethylsiloxane) and conductive carbon nanotubes/graphene network. *ACS Appl. Mater. Interfaces* **2016**, *8*, 2187–2192. [CrossRef]
20. Zhang, K.; Shi, X.; Chen, J.; Xiong, T.; Jiang, B.; Huang, Y. Self-healing and stretchable PDMS-based bifunctional sensor enabled by synergistic dynamic interactions. *Chem. Eng. J.* **2021**, *412*, 128734. [CrossRef]
21. Martinez, V.; Stauffer, F.; Adagunodo, M.O.; Forro, C.; Vörös, J.; Larmagnac, A. Stretchable silver nanowire–elastomer composite microelectrodes with tailored electrical properties. *ACS Appl. Mater. Interfaces* **2015**, *7*, 13467–13475. [CrossRef] [PubMed]
22. Feng, P.; Zhong, M.; Zhao, W. Stretchable multifunctional dielectric nanocomposites based on polydimethylsiloxane mixed with metal nanoparticles. *Mater. Res. Express* **2019**, *7*, 015007. [CrossRef]
23. Zou, Q.; He, K.; Ou-Yang, J.; Zhang, Y.; Shen, Y.; Jin, C. Highly Sensitive and Durable Sea-Urchin-Shaped Silver Nanoparticles Strain Sensors for Human-Activity Monitoring. *ACS Appl. Mater. Interfaces* **2021**, *13*, 14479–14488. [CrossRef] [PubMed]
24. Al-Milaji, K.N.; Huang, Q.; Li, Z.; Ng, T.N.; Zhao, H. Direct embedment and alignment of silver nanowires by inkjet printing for stretchable conductors. *ACS Appl. Electron. Mater.* **2020**, *2*, 3289–3298. [CrossRef]

25. Duan, S.; Wang, Z.; Zhang, L.; Liu, J.; Li, C. Three-dimensional highly stretchable conductors from elastic fiber mat with conductive polymer coating. *ACS Appl. Mater. Interfaces* **2017**, *9*, 30772–30778. [CrossRef]
26. Zhang, S.; Li, Y.; Tian, Q.; Liu, L.; Yao, W.; Chi, C.; Wu, W. Highly conductive, flexible and stretchable conductors based on fractal silver nanostructures. *J. Mater. Chem. C* **2018**, *6*, 3999–4006. [CrossRef]
27. Zhou, W.; Yu, Y.; Bai, S.; Hu, A. Laser direct writing of waterproof sensors inside flexible substrates for wearable electronics. *Opt. Laser Technol.* **2021**, *135*, 106694. [CrossRef]
28. Chou, N.; Kim, Y.; Kim, S. A method to pattern silver nanowires directly on wafer-scale PDMS substrate and its applications. *ACS Appl. Mater. Interfaces* **2016**, *8*, 6269–6276. [CrossRef]
29. Sun, J.; Wang, Q.; Luo, G.; Meng, W.; Cao, M.; Li, Y.; Lang, M.F. A novel flexible Ag/AgCl quasi-reference electrode based on silver nanowires toward ultracomfortable electrophysiology and sensitive electrochemical glucose detection. *J. Mater. Res. Technol.* **2020**, *9*, 13425–13433. [CrossRef]
30. Hu, J.; Yu, J.; Li, Y.; Liao, X.; Yan, X.; Li, L. Nano carbon black-based high-performance wearable pressure sensors. *Nanomaterials* **2020**, *10*, 664. [CrossRef]
31. Zhao, C.; Xia, Z.; Wang, X.; Nie, J.; Huang, P.; Zhao, S. 3D-printed highly stable flexible strain sensor based on silver-coated-glass fiber-filled conductive silicon rubber. *Mater. Des.* **2020**, *193*, 108788. [CrossRef]
32. Choi, S.; Kim, S.; Kim, H.; Lee, B.; Kim, T.; Hong, Y. 2-D Strain Sensors Implemented on Asymmetrically Bi-Axially Pre-Strained PDMS for Selectively Switching Stretchable Light-Emitting Device Arrays. *IEEE Sens. J.* **2020**, *20*, 14655–14661. [CrossRef]
33. Xiang, D.; Zhang, X.; Harkin-Jones, E.; Zhu, W.; Zhou, Z.; Shen, Y.; Wang, P. Synergistic effects of hybrid conductive nanofillers on the performance of 3D printed highly elastic strain sensors. *Compos. Part A Appl. Sci. Manuf.* **2020**, *129*, 105730. [CrossRef]

crystals

Article

Dye Sequestration Using Biosynthesized Silver Nanoparticles Adsorbent in Aqueous Solutions

Madiha Batool [1], Walid M. Daoush [2,3,*] and Muhammad Khalid Hussain [4]

1 Department of Chemistry, GC University, Lahore 54000, Pakistan; tweetchem56@gmail.com
2 Department of Production Technology, Faculty of Technology and Education, Helwan University, Cairo 11281, Egypt
3 Department of Chemistry, College of Science, Imam Mohammad Ibn Saud Islamic University (IMSIU), Riyadh 11623, Saudi Arabia
4 Department of Physics, Hafiz Hayat Campus, University of Gujrat, Gujrat 50700, Pakistan; muhammad.khalid@uog.edu.pk
* Correspondence: wmdaoush@imamu.edu.sa

Abstract: Nanomaterials have gained much attention in the field of environmental remediation, largely due to their high surface area-to-volume ratio and other unique physical, chemical, and biological characteristics that emerge due to its size effects. Metallic nanoparticles are traditionally manufactured using wet chemical processes; however, the chemicals utilized are generally hazardous and combustible. The biosynthesis of nanoparticles using a variety of plant resources is considered a green technology because it does not use toxic chemicals. This work focuses on the green synthesis of biogenetic silver nanoparticles and their use in the sequestration of colorants from aqueous solution. The extract of aquatic macrophyte *Salvinia molesta* (water hyacinth) has been employed to prepare silver nanoparticles by chemical reduction reaction. In the UV-visible spectrum of the synthesized silver nanoparticles, the absorbance peak was detected in the 420–430 nm range. The synthesized silver nanoparticles were used to sequester methylene blue (MB) dye in aqueous solution. About 121.04 mg/g was found as the highest adsorption capacity of methylene blue dye on the silver nanoparticles according to the Langmuir isotherm. It was observed that the experimental results and the pseudo-second order kinetics are in good agreement. As a result, the biosynthesized silver nanoparticle might be a potential adsorbent material in the field of environmental rehabilitation and cleanup.

Keywords: silver nanoparticles; green synthesis; adsorption; colorants; methylene blue dye

Citation: Batool, M.; Daoush, W.M.; Hussain, M.K. Dye Sequestration Using Biosynthesized Silver Nanoparticles Adsorbent in Aqueous Solutions. *Crystals* **2022**, *12*, 662. https://doi.org/10.3390/cryst12050662

Academic Editor: Witold Łojkowski

Received: 15 March 2022
Accepted: 29 April 2022
Published: 5 May 2022

1. Introduction

Textile production requires a large quantity of water and energy. Factories create a large amount of contaminated water after finishing dyeing and completing their products [1]. Due to its huge volume and nature, this water is regarded to be extremely dangerous for human beings. There are a number of environmental problems associated with discharging dyeing waste water directly into the environment. Waste water has a potential effect to cause serious damage to aquatic life, soil, and drinking water [2]. Furthermore, some dyes and their tailing-remained products have the potential to be carcinogenic and poisonous. As a result, the dyes must be removed before disposal. A wide range of techniques have been explored to treat wastewater, including biodegradation and ultrafiltration as well as photocatalytic degradation, oxygenation, and adhesion. When it comes to the treatment of these dye-containing effluents, traditional adsorption is a very low-cost and effective method [3]. The physicochemical characteristics of the adsorbent determine the efficiency of any adsorption procedure. The search for novel adsorbents with large specific surface area, high adsorption capacity, and a quick adsorption rate, as well as unique surface reactivity, is therefore highly essential and useful [4].

Nanotechnology allows the creation of nanomaterials that may be used for a variety of scientific and technical purposes. As a result of nanotechnology, new nanomaterials may be used for a variety of applications with a measurement of less than 100 nanometers [5]. Research on nanoparticles has received increased attention in recent years, owing to a wide range of essential uses in health and different environmental issues. As a result of their tiny size, nanoparticles have a higher surface-to-volume ratio or a greater surface area per weight than larger particles, making them more reactive to interact with other molecules. Silver nanoparticles (AgNPs) are more important than other noble metal nanoparticles due to their specific features, such as excellent optical property, electrical conductivity, oxidative property, antibacterial activity, and catalytic activity. Silver nanoparticles are traditionally produced using hazardous chemicals and at a high cost. The utilization of environmentally friendly resources such as plant leaf extract, microorganisms, and enzymes may be employed for the biological production of nanoparticles, which offers several environmental benefits. The green production of nanoparticles is critical for the treatment of many hazardous compounds that cannot be addressed chemically [6]. The fabrication of nanoparticles using chemical methods are time-consuming and have a high production cost. Therefore, scientists were particularly interested in the creation of nanoparticles by green production methods.

In the literature, researchers fabricated silver nanoparticles and applied for different applications. Vanaja et al. (2014) prepared silver nanoparticles using *Morinda tinctoria* leaf extract under specific pH conditions and further characterized by different spectroscopic techniques [7]. They concluded that the size as well as the quantity of silver nanoparticles formed are strongly dependent on the pH, and basic pH supports the biosynthesis of silver nanoparticles, whereas no silver nanoparticles were detected in the acid medium. Silver nanoparticles were prepared with sizes ranging from 79 to 90 nm and applied for photo catalytic activity of MB degradation [8]. Anu Kumar et al. (2016) experimented with a simple and non-hazardous method for synthesizing silver Nano catalyst by using *Viola serpens* leaf extract [9].

The biosynthesis of AgNPs mainly involves three main steps: (1) solvent medium should be selected, (2) environmentally friendly reducing agents are selected, and (3) nontoxic substances are chosen for the AgNPs stability. When a substrate of low value is used by living cells to produce higher-value products, this device is called a bioreactor [10] Plant parts act as a bioreactor for the expression in tissues remote from the penetration of ion sites, with the capability to reduce metal ions on the surface. The plant has a strong hyperaccumulation capacity of metal ions, which is a tool for the bioaccumulation of metal oxide-NPs. The whole plant can be used for nanoparticle fabrication, whereas size and shape of NPs mainly depends upon parts of the plant. Biogenic fabrication by plant extracts have been used in the production of AgNPs and found interesting owing to size, surface area, structure, and unique properties. There are several disadvantages of physical and chemical methods of nanometal fabrication, as they need to use biogenic synthesis methods, which are economically feasible and ecofriendly [11]. The purpose of this work is to synthesize AgNP using the aquatic weed, *Salvinia molesta*, as a reducing agent and assess their effectiveness in sequencing MB dye, which is commonly used in the textile industry. It was shown that the synthesized AgNPs by plant extracts adsorb and quickly decrease the concentrations of the MB dye, both in terms of catalysis and kinetics [12].

2. Materials and Methods

2.1. Materials

Silver nitrate and methylene blue (MB) dye were purchased from Sigma Aldrich. *Salvinia molesta*, a macrophyte kariba weed, is an aquatic fern and native to southeastern Brazil [13]. Its observed as a free-floating plant that remains on the surface of water. The leaves are 0.1–4 cm long, with a bristly surface caused by the hair-like strands and were collected from riverbanks and ponds in the summer of 2021 (see Figure 1).

Figure 1. Optical images with (**a**) low and (**b**) high magnifications of the Aquatic macrophyte (*Salvinia molesta*) leaves under investigation.

2.2. Methods

2.2.1. Plant Extraction

The collected *S. molesta* was washed many times in warm distilled water to remove any dust or other impurities. A total of 50 g of *S. molesta* leaves were cleaned and finely chopped before being placed in 500 mL beakers containing 200 mL of double-distilled water. The mixture was allowed to boil for 1 h at 100 °C. The obtained solution mixture was then cooled to room temperature [14]. The leaf extracts were filtered by using Whatman number 41 filter paper.

2.2.2. Silver Nanoparticle (AgNPs) Synthesis

In order to synthesize AgNPs, an aqueous solution of silver nitrate was prepared using double-distilled water. AgNPs were made by mixing 1 mM silver nitrate solution with *S. molesta* leaf extract and continuous stirring at 300 rpm for 72 h at room temperature using a magnetic stirrer. The formation of AgNPs is indicated by the appearance of a yellow-brown color after 8 h of agitation and turns reddish brown after 72 h of agitation in the dark [15,16]. The resulting suspension of silver nitrate and *S. molesta* extract were centrifuged at 15,000 rpm for 15 min. To remove silver ions and seed extract residue, the pellet containing AgNPs was washed three to four times with deionized water. Biosynthesized AgNPs were lyophilized after being precipitated. To further characterize, the lyophilized nanoparticles were stored in a cool, dry, and dark environment.

2.2.3. Characterization of Biogenic AgNPs

The mixture of silver nitrate and *S. molesta* extract solution were taken in different ratios (1:1, 1:2, 1:3, and 1:4), and the UV-visible spectra of AgNPs at different ratios were obtained using a UV-1800 Shimadzu spectrophotometer. Silver nanoparticle morphology was studied using a lyophilized sample of AgNPs using a scanning electron microscope (SEM) operating at 20 kV. Transmission electron microscopy (TEM) working at an acceleration voltage of 200 kV was used to determine the morphology and size of biosynthesized AgNPs. The crystalline structure of the biogenic nanoparticles was determined using a pattern of selected area electron diffraction (SAED).

2.2.4. Dye Adsorption and Sequestration Using AgNPs Adsorbate

For the sequestration of MB dye, batch studies were carried out in a 250 mL conical flask containing 50 mL of dye solution. The effects of initial dye concentration (10–100 mg/L), equilibrium time, pH (2–10), and temperature (30–45 °C) were investigated. After adding a desired amount (0.1 g) of AgNPs to MB dye solution, the mixture was stirred at 200 rpm with a magnetic stirrer at an ambient temperature for equilibrium time. After equilibrium time, the solution was filtered to separate pellet and supernatant [17–21]. The collected supernatant was analyzed using UV-Vis spectrophotometer at a maximum wavelength of 660 nm to measure the MB dye concentration in the residual solution. The efficiency of decolorization (%) has been calculated by using the following Equation (1) [10]:

$$\text{Decolorization}(\%) = \frac{C_0 - C_1}{C_0} \times 100\% \quad (1)$$

where C_0 is the initial concentration of dye and C_1 is the concentration of dye after irradiation in the selected time interval.

3. Results and Discussion

3.1. *Characterization of Biosynthesized AgNPs*

Silver nanoparticles were synthesized by the addition of Salvinia molesta extract to the $AgNO_3$ solution. In the $AgNO_3$/Salvinia molesta solution, silver nanoparticle synthesis is indicated by a progressive color change from colorless to yellow to reddish brown, as shown in Figure 2. Surface plasmon vibration, an optical characteristic found only in noble metals, is responsible for this color change [22]. Further, UV-visible spectroscopy has been used to confirm the formation of silver nanoparticles in aqueous solution. The solution was scanned over a range of wavelength as 300–800 nm. Biogenic AgNPs have a prominent peak absorbance at 430 nm [23], which is typical, as shown in Figure 3. *Salvinia molesta* leaf extract generated highly dense AgNPs, as shown by SEM examination in Figure 4. The acquired morphology demonstrated that the produced AgNPs were virtually spherical. The particles appear agglomerated because the presence of several significant bio-organic chemicals extruded from the leaves appear as a chelating agent that stabilizes the produced AgNPs in solution. A similar phenomenon was reported by Ranjith kumar et al. (2018) [24].

8 h 16 h 24 h 48 h 72 h

Figure 2. Color change during AgNPs biosynthesis at different time intervals.

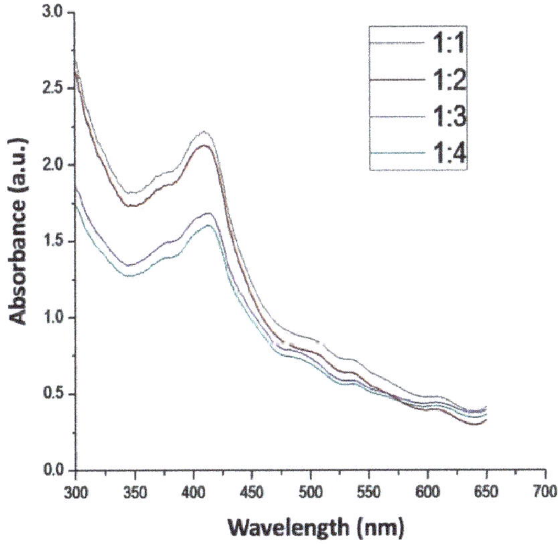

Figure 3. UV spectra of the biosynthesized AgNPs using different ratios of silver nitrate to *S. molesta* leaves extract.

A High-resolution transmission electron microscopy (HRTEM) investigation was carried out to characterize the surface morphology and size (Figure 5a–c). The average particle size was determined to be 1 nm. Additionally, the particles were rounded and well-dispersed, suggesting the presence of capping peptides around each particle that help stabilize the nanoparticles. For a face-centered cubic structure, the SAED patterns recorded for a single particle in the aggregates corresponded to a characteristic polycrystalline ring pattern as shown in the Figure 5d [25]. The bright circular ring observed is due to the reflection from the lattice planes of crystalline biosynthesized AgNPs.

Figure 4. SEM image of the biosynthesized AgNPs.

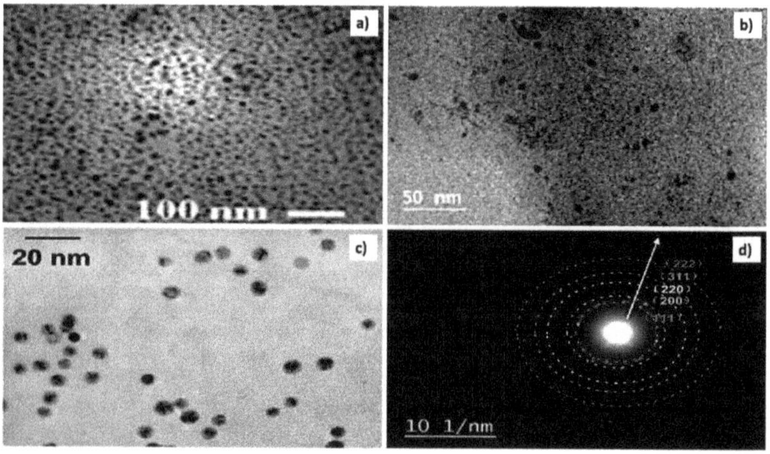

Figure 5. (**a–c**) TEM image and (**d**) SAED diffraction pattern of the biosynthesized AgNPs.

3.2. *Effect of Dye Concentration on the Adsorption Process*

Adsorption processes are depending on the initial concentration of adsorbate. Initially, a large number of surface-active sites will be available for adsorbing the desired dye molecules. So, the adsorption is fast at the initial stage. Figure 6 shows the effect of time and concentration of the investigated MB dye with respect to decolorization percentage. A decrease in degradation efficiency occurs as dye concentration increases due to several factors. As there are fewer active sites on the surface, hydroxyl radicals will be less active, and the probability of a photon reaching the surface of AgNPs is decreased with increasing the dye concentration, resulting in a decreased decolorization percentage [26].

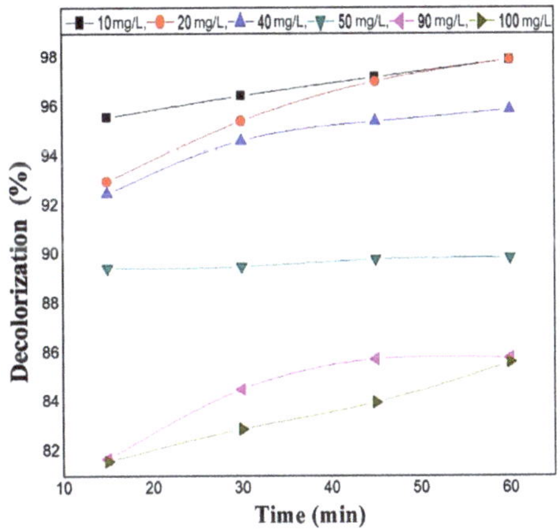

Figure 6. Effect of time on the decolorization percentage of methylene blue dye adsorbed on the biosynthesized AgNPs.

3.3. Effect of pH on the Adsorption Process

Each of the adsorbate solutions were prepared at different pH levels (2.0, 4.0, 6.0, 8.0, and 10) using either 1.0 N HCl or 1.0 N NaOH. There were optimum amounts of adsorbents in dye solutions that were added (0.1 g of AgNPs), and the mixture was agitated. The amount of dye that was adsorbed was measured. From the plot of pH versus percentage of dye removal (Figure 7), it was observed that the optimum pH of the process was 4. This is due to the fact that the degradation of MB dye was increased. The neutralization of the negative charges at the surface of AgNPs will enhance the diffusion phenomena to occur, and more active sites will be available for the adsorption process. However, when the pH is below 4, the dye degradation and the number of negatively charged adsorbent sites decreases. The increasing in the number of positively charged surface active sites enhances the dye-adsorbent repulsion, which reduces the efficiency of the adsorption process. At a lower pH, the MB adsorption dye improved due to the acidic media. As pH increases, the adsorbent surface acquired negatively charges, and they are responsible for increasing the adsorption capacity due to the electrostatic interaction. However, the adsorption of MB dye was observed decreasing at high pH.

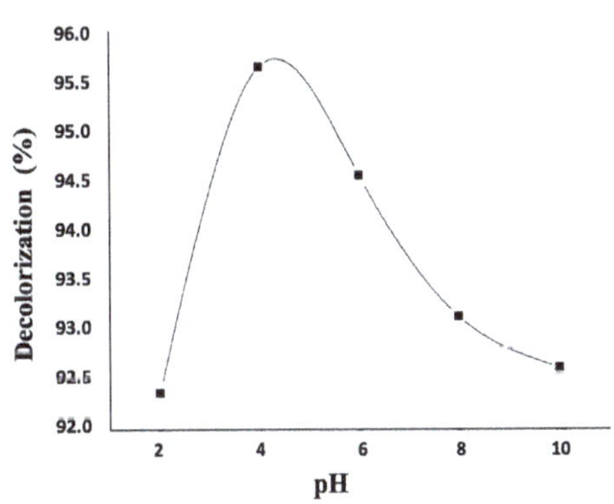

Figure 7. Effect of pH on the decolorization percentage of the methylene blue dye adsorbed on the biosynthesized AgNPs.

3.4. Effect Temperature on the Adsorption Process

An initial MB dye concentration of 10 mg/L was used to determine the effect of temperature on degradation efficiency of MB on the surface of the biosynthesized AgNPs. Figure 8 exhibits the influence of temperature on percentage decolorization. According to the study, the optimal temperature for dye degradation was found to be 35 °C. Further, when the temperature was raised above 35 °C, the dye adsorption decreased significantly. This is due to the fact that the adsorptive forces at the adsorbent active sites and the adsorbate may be weakened.

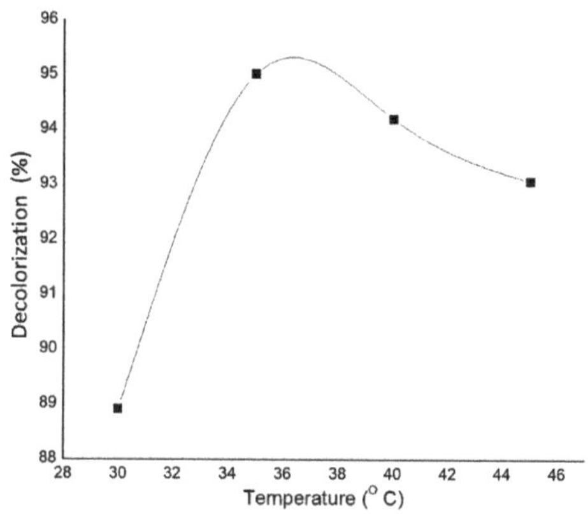

Figure 8. Effect of temperature on the decolorization percentage of the methylene blue dye adsorbed on the biosynthesized AgNPs.

3.5. Adsorption Isotherms

The interaction between adsorbate and adsorbent is described using adsorption isotherms. Langmuir and Freundlich isotherm models are used to assess experimental and theoretical MB dye adsorption data. According to Langmuir, adsorption occurs uniformly or homogeneously on the surface of the biosynthesized AgNPs adsorbent. Langmuir's linear expression can be found in Equation (2) [10];

$$\frac{C_e}{q_e} = \frac{1}{Qb} + \frac{C_e}{Q} \tag{2}$$

where C_e is the equilibrium concentration of a dye in solution (mg/L), q_e is the amount of dye adsorbed on to the Cs (mg/g) MB, Q is the Langmuir constant related to adsorption capacity (mg/g), and b is the Langmuir constant related to sorption energy (L/mg). The experimental data for the decolorization of MB dye is fitted in the rearranged Langmuir equation, which has found successful applications of monolayer adsorption. The plot obtained is shown in Figure 9. A Freundlich isotherm is used to determine the adsorption capacity of MB dye on the biosynthesized AgNPs. Equation (3) is the linear form of the Freundlich isotherm.

$$\log q_e = \log k_f + \frac{1}{n} \log C_e \tag{3}$$

where C_e is the equilibrium concentration of the dye in solution (mg/L), q_e is the amount of dye adsorbed on the adsorbent (mg/g), and K_f and $1/n$ are Freundlich constants [27]. The graph of log q_e is plotted against log C_e and is found not to be linear as shown in Figure 10. The $1/n$ value from the experimental data is greater than 1, which shows unsatisfactory adsorption of MB dye onto the surface of the biosynthesized AgNPs adsorbent. The parameters of Langmuir and Freundlich constants are given in Table 1. The Langmuir isotherm was found to be the best fit in the experimental data—better than the Freundlich isotherm. This shows the monolayer adsorption of MB dye onto the biosynthesized AgNPs adsorbent, with the maximum adsorption capacity of 1.023 mg/g.

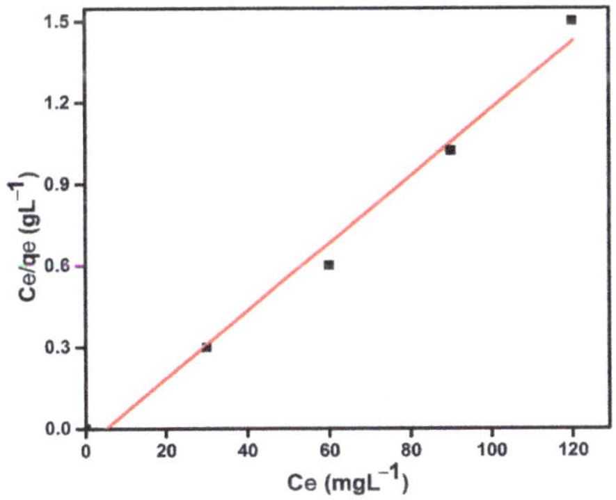

Figure 9. Decolorization percentage of the methylene blue dye as adsorbed on the biosynthesized AgNPs. (Fitted with the Langmuir model).

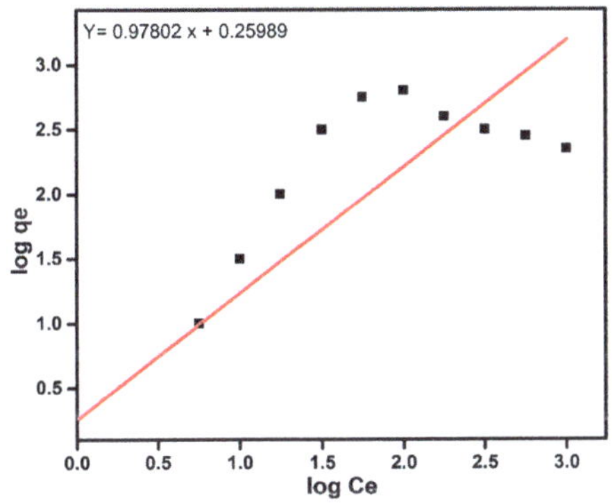

Figure 10. Relationship between log q_e with log C_e of the adsorbed methylene blue dye on the biosynthesized AgNPs.

Table 1. Freundlich and Langmuir isotherm parameters used at the adsorption of methylene blue dye onto the surface of the biosynthesized AgNPs adsorbent.

Freundlich Parameters			Langmuir Parameters		
k_f	n	R^2	Q_m (mg/g)	b (L/mg)	R^2
1.0023	1.100	0.931	121.04	0.2842	0.998

3.6. Adsorption Kinetics

For the study of adsorption kinetics, pseudo first order and pseudo second order kinetic models are widely used. These models are used to determine the rate at which adsorption occurs or the rate at which solute is absorbed. When it comes to designing the reactors, these models pay more attention. Our study examined the rate at which MB dye adsorbed onto surfaces by applying pseudo first- and second-order equations in order to determine how fast it adsorbed. There is a strong correlation between experimental data and predicted values using different models (value close or equal to 1).

3.6.1. Pseudo First Order Kinetic Model

The rate of MB dye adsorption on the biosynthesized AgNPs adsorbent's surface is proportional to the amount of dye adsorbed from the liquid phase; the pseudo first order kinetic equation may be expressed as in Equation (4) [10]:

$$\log q_e - q_t = \log q_e - \frac{k_{ad}t}{2.303} \quad (4)$$

where q and q_e represent the amount of dye adsorbed (mg/g) at time t and at equilibrium time, respectively, where k_{ad} is the adsorption rate constant [28]. Figure 11 shows the plot of linearized form of the pseudo first order kinetic model. The slopes and intercepts of the plotted graph of (log $q_e - q_t$) versus time were used to determine the pseudo first order rate constant k_{ad} and the equilibrium adsorption capacity q_e.

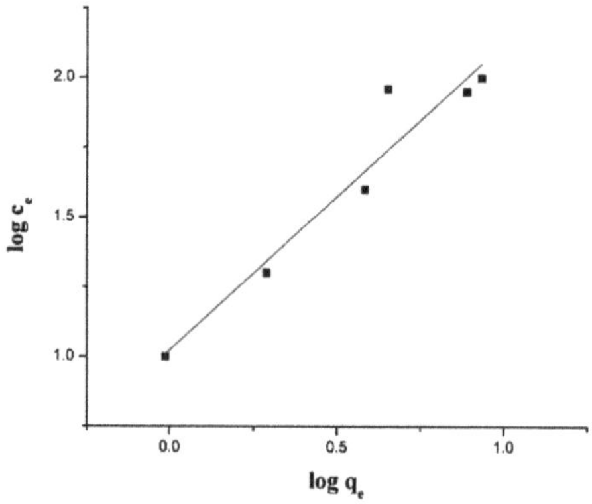

Figure 11. Pseudo first order kinetic model of the adsorption process of methylene blue dye on the biosynthesized AgNPs.

3.6.2. Pseudo Second Order Kinetic Model

The pseudo second order kinetics model can be expressed by Equation (5) [10]:

$$dq_t/qt = k_2(q_e - q_t)^2 \quad (5)$$

where k_2 represents the pseudo second order rate constant. The q_e and q_t represent the amount of dyes adsorbed (mg/g) at equilibrium and at time t. For the boundary condition $t = 0$ to $t = t$ and $q_t = 0$ to $q_t = t$, the integral form of the Equation (3) becomes Equation (6) [10]:

$$t/q_t = 1/k_2 q_e^2 + t/q_e \quad (6)$$

The values of the k_2 and q_e are calculated from the intercepts and slopes of the plots of t/q_t vs. time (Figure 12), and the corresponding coefficient of determined R^2 values are listed in Table 2. The experimental value of q_e is 10 mg/L and the theoretical value is 9.708 mg/L for second order. From this reason, it can be interpreted that the adsorption process of methylene blue dye on the biosynthesized AgNPs follows second order kinetics [29–31].

Table 2. Pseudo first and second order kinetic models.

Pseudo First Order			Pseudo Second Order		
q_e (mg/g)	k_{ad} (1/min)	R^2	q_e (mg/g)	k_{ad} (g/mg/min)	R^2
81.01	0.046	0.916	97.08	0.707	0.99

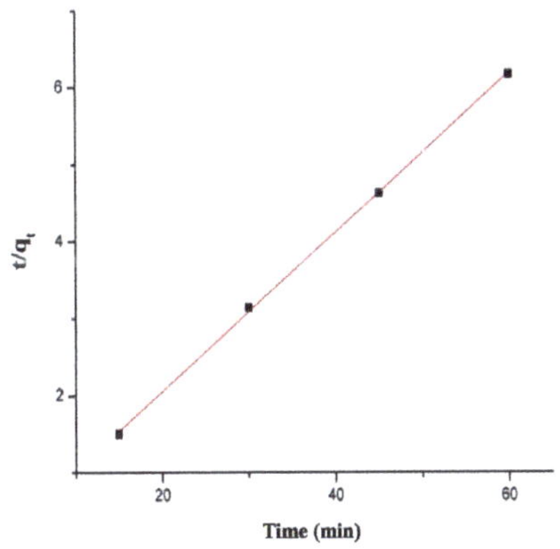

Figure 12. Pseudo second order Kinetic model—the adsorption process of the methylene blue dye on the biosynthesized AgNPs.

As a result of these findings, the prepared nanoparticles mediated by *S. molesta* could be used effectively in environmental remediation, and their phenomenal behavior might be extensively utilized to breakdown hazardous organic colorants from different industrial effluents. The pseudo first order (P.F.O) reaction R^2 value was calculated as 0.91 as compared to pseudo second order (P.S.O) reaction 0.99 and fitted well. Decolorization percentage of methylene blue dye as adsorbed on the AgNPs was fitted with the Langmuir model.

4. Conclusions

This study highlights the use of nanotechnology for methylene blue dye adsorption on biosynthesized silver nanoparticles with effective results. It was shown that bioactive components of the aquatic macrophyte (*S. molesta*) extracts can be used as a reducing agent to fabricate silver nanoparticles used efficiently in methylene blue dye adsorption. This is a quick method for the production of well-defined silver nanoparticles, as evidenced by UV-Vis, TEM, SEM, and SAED methods. The efficient activity of silver nanoparticles enhancing the methylene blue dye degradation was observed. The biosynthesized Ag-NPs have high adsorption activity against the decolorization of methylene blue dye. The present study could be an indication for further future research to identify other nanoparticles for different synthetic dyes for efficient degradation potential.

Author Contributions: Conceptualization M.B., W.M.D. and M.K.H.; Methodology, M.B., W.M.D. and M.K.H.; formal analysis, M.B., W.M.D. and M.K.H.; investigation, M.B., W.M.D. and M.K.H.; resources, M.B., W.M.D. and M.K.H.; writing original draft preparation, M.B., W.M.D. and M.K.H.; writing review and editing, M.B., W.M.D. and M.K.H. All authors have read and agreed to the published version of the manuscript.

Funding: This research received no external funding.

Institutional Review Board Statement: Not applicable.

Informed Consent Statement: Not applicable.

Data Availability Statement: All data are available in this manuscript.

Acknowledgments: The authors are thankful to the staff of the METS School of Engineering, Thrissur, India, for their cooperation during conducting this research work.

Conflicts of Interest: The authors declare no conflict of interest.

References

1. Aboelfetoh, E.F.; Gemeay, A.H.; El-Sharkawy, R.G. Effective disposal of methylene blue using green immobilized silver nanoparticles on graphene oxide and reduced graphene oxide sheets through one-pot synthesis. *Environ. Monit. Assess.* **2020**, *192*, 1–20. [CrossRef]
2. Hussain, M.K.; Khalid, N.R. Surfactant-assisted synthesis of MoO_3 nanorods and its application in photocatalytic degradation of different dyes in aqueous environment. *J. Mol. Liq.* **2022**, *346*, 117871. [CrossRef]
3. Wenjie, Z.; Zhou, C.; Zhou, W.; Lei, A.; Zhang, Q.; Wan, Q.; Zou, B. Fast and considerable adsorption of methylene blue dye onto graphene oxide. *Bull. Environ. Contam. Toxicol.* **2011**, *87*, 86–90.
4. Khalid, N.R.; Hussain, M.K.; Murtaza, G.; Ikram, M.; Ahmad, M.; Hammad, A. A Novel $Ag_2O/Fe–TiO_2$ Photocatalyst for CO_2 Conversion into Methane Under Visible Light. *J. Inorg. Organomet. Polym. Mater.* **2019**, *29*, 1288–1296. [CrossRef]
5. Ho, Y.S.; Chiang, C.C. Sorption studies of acid dye by mixed sorbents. *Adsorption* **2001**, *7*, 139–147. [CrossRef]
6. Bar, H.; Bhui, D.K.; Sahoo, G.P.; Sarkar, P.; De, S.P.; Misra, A. Green synthesis of silver nanoparticles using latex of *Jatropha curcas*. *Colloids Surf. A Physicochem. Eng. Asp.* **2009**, *339*, 134–139. [CrossRef]
7. Yang, Z.; Zhang, T.; Ren, J.; Li, J.; Ge, J.; Shan, H.; Ji, T.; Xu, M.; Liu, Q. Nano-silver functionalized spherical activated carbon with enhanced dipropyl sulfide adsorption capacity and antibacterial properties. *RSC Adv.* **2022**, *12*, 9933–9943. [CrossRef] [PubMed]
8. Liu, Y.; Shi, T.; Si, Q.; Liu, T. Adsorption and sensing performances of transition metal (Pd, Pt, Ag and Au) doped MoTe2 monolayer upon NO_2: A DFT study. *Phys. Lett. A* **2021**, *391*, 127117. [CrossRef]
9. Rezaei, A.; Rezaei, M.R.; Sayadi, M.H. 3D network structure graphene hydrogel-Fe_3O_4@ SnO_2/Ag via an adsorption/photocatalysis synergy for removal of 2,4 dichlorophenol. *J. Taiwan Inst. Chem. Eng.* **2021**, *121*, 154–167. [CrossRef]
10. Ramalingam, B.; Khan, M.M.R.; Mondal, B.; Mandal, A.B.; Das, S.K. Facile synthesis of silver nanoparticles decorated magnetic-chitosan microsphere for efficient removal of dyes and microbial contaminants. *ACS Sustain. Chem. Eng.* **2015**, *3*, 2291–2302. [CrossRef]
11. Iravani, S.; Korbekandi, H.; Mirmohammadi, S.V.; Zolfaghari, B. Synthesis of silver nanoparticles: Chemical, physical and biological methods. *Res. Pharm. Sci.* **2014**, *9*, 385. [PubMed]
12. Ponarulselvam, S.; Panneerselvam, C.; Murugan, K.; Aarthi, N.; Kalimuthu, K.; Thangamani, S. Synthesis of silver nanoparticles using leaves of *Catharanthus roseus* Linn. G. Don and their antiplasmodial activities. *Asian Pac. J. Trop. Biomed.* **2012**, *2*, 574–580. [CrossRef]
13. Sivashankar, R.; Susheeba, O.K.; Sivasubramanian, V. Adsorption of organic dye on to novel magnetic biocomposite: Kinetics and equilibrium studies. *Res. J. Chem. Environ.* **2015**, *19*, 48–57.
14. Ibrahim, H.M. Green synthesis and characterization of silver nanoparticles using banana peel extract and their antimicrobial activity against representative microorganisms. *J. Radiat. Res. Appl. Sci.* **2015**, *8*, 265–275. [CrossRef]
15. Saha, N.; Trivedi, P.; Gupta, S.D. Surface plasmon resonance (SPR) based optimization of biosynthesis of silver nanoparticles from rhizome extract of *Curculigo orchioides* Gaertn. and its antioxidant potential. *J. Clust. Sci.* **2016**, *27*, 1893–1912. [CrossRef]
16. Rajamani, R.K.; Kuppusamy, S.; Bellan, C. Biosynthesis, characterization and remedical aspect of silver nanoparticles against pathogenic bacteria. *MOJ Toxicol.* **2018**, *4*, 103–109. [CrossRef]
17. Nam, W.; Kim, J.; Han, G. Photocatalytic oxidation of methyl orange in a three-phase fluidized bed reactor. *Chemosphere* **2002**, *47*, 1019–1024. [CrossRef]
18. Banerjee, S.; Chattopadhyaya, M.C. Adsorption characteristics for the removal of a toxic dye, tartrazine from aqueous solutions by a low-cost agricultural by-product. *Arab. J. Chem.* **2017**, *10*, S1629–S1638. [CrossRef]
19. Ünal, G.; Özcan, G.; Gürpınar, G.Ç. Removal of methylene blue from aqueous solution by activated carbon prepared from pea shells (*Pisum sativum*). *J. Chem.* **2013**, *2013*, 614083.
20. Ghaedi, M.; Ansari, A.; Habibi, M.H.; Asghari, A.R. Removal of malachite green from aqueous solution by zinc oxide nanoparticle loaded on activated carbon: Kinetics and isotherm study. *J. Ind. Eng. Chem.* **2014**, *20*, 17–28. [CrossRef]
21. Khalid, N.R.; Hammad, A.; Tahir, M.B.; Rafique, M.; Iqbal, T.; Nabi, G.; Hussain, M.K. Enhanced photocatalytic activity of Al and Fe co-doped ZnO nanorods for methylene blue degradation. *Ceram. Int.* **2019**, *45*, 21430–21435. [CrossRef]
22. Hussain, M.K.; Khalid, N.R.; Tanveer, M.; Kebaili, I.; Alrobei, H. Fabrication of CuO/MoO_3 pn heterojunction for enhanced dyes degradation and hydrogen production from water splitting. *Int. J. Hydrogen Energy* **2022**, *47*, 15491–15504. [CrossRef]
23. Naghizade Asl, M.; Mahmodi, N.M.; Teymouri, P.; Shahmoradi, B.; Rezaee, R.; Maleki, A. Adsorption of organic dyes using copper oxide nanoparticles: Isotherm and kinetic studies. *Desalin. Water Treat.* **2016**, *57*, 25278–25287. [CrossRef]
24. Alver, E.; Ayşegül, Ü.M. Anionic dye removal from aqueous solutions using modified zeolite: Adsorption kinetics and isotherm studies. *Chem. Eng. J.* **2012**, *200*, 59–67. [CrossRef]
25. Khalid, N.R.; Arshad, A.; Tahir, M.B.; Hussain, M.K. Fabrication of p–n heterojunction Ag_2O@ Ce_2O nanocomposites make enables to improve photocatalytic activity under visible light. *Appl. Nanosci.* **2021**, *11*, 199–206. [CrossRef]
26. Agarwal, H.; Shanmugam, V.K. A review on anti-inflammatory activity of green synthesized Zinc Oxide nanoparticle: Mechanism-based approach. *Bioorg. Chem.* **2020**, *94*, 103423. [CrossRef]
27. Yadav, R.K.; Singh, A.K.; Singh, N.B.; Singh, N.; Khare, S.; Pandey, A.K. Green Synthesized Nanoparticle-Mediated Wastewater Treatment. In *Emerging Eco-Friendly Green Technologies for Wastewater Treatment*; Springer: Singapore, 2020; pp. 299–309.
28. Das, M.; Samal, A.K.; Mhar, N. Butea monosperma leaf as an adsorbent of methylene blue: Recovery of the dye and reuse of the adsorbent. *Int. J. Environ. Sci. Technol.* **2020**, *17*, 2105–2112. [CrossRef]

29. Vizuete, K.S.; Kumar, B.; Vaca, A.V.; Debut, A.; Cumbal, L. Mortiño (*Vaccinium floribundum Kunth*) berry assisted green synthesis and photocatalytic performance of Silver–Graphene nanocomposite. *J. Photochem. Photobiol. A Chem.* **2016**, *329*, 273–279. [CrossRef]
30. Vanaja, M.; Paulkumar, K.; Baburaja, M.; Rajeshkumar, S.; Gnanajobitha, G.; Malarkodi, C.; Sivakavinesan, M.; Annadurai, G. Degradation of Methylene Blue Using Biologically Synthesized Silver Nanoparticles. *Bioinorg. Chem. Appl.* **2014**, *2014*, 742346. [CrossRef]
31. Kumar, A.; Mazumdar, R.S.; Dhewa, T. Biological synthesis of silver nanoparticles by using *Viola serpens* extract. *Asian Pac. J. Trop. Dis.* **2016**, *6*, 223–226. [CrossRef]

crystals

MDPI

Article

The Impact of Laminations on the Mechanical Strength of Carbon-Fiber Composites for Prosthetic Foot Fabrication

Bakhtawar Sehar [1], Asim Waris [1], Syed Omer Gilani [1], Umar Ansari [1], Shafaq Mushtaq [2], Niaz B. Khan [1], Mohammed Jameel [3], M. Ijaz Khan [4,5,*], Omar T. Bafakeeh [6] and El Sayed Mohamed Tag-ElDin [7]

1 Department of Biomedical Engineering and Sciences, School of Mechanical and Manufacturing Engineering, National University of Sciences and Technology (NUST), Islamabad 44000, Pakistan
2 Pakistan Institute of Medical Sciences G-8/3, Islamabad 44000, Pakistan
3 Department of Civil Engineering, King Khalid University, Abha 61421, Saudi Arabia
4 Department of Mechanical Engineering, Lebanese American University, Beirut P.O. Box 36, Lebanon
5 Department of Mathematics and Statistics, Riphah International University I-14, Islamabad 44000, Pakistan
6 Department of Engineering Sciences, Jazan University, Jazan 45124, Saudi Arabia
7 School of Engineering, Future University in Egypt, New Cairo 118935, Egypt
* Correspondence: ijazfmg_khan@yahoo.com or mikhan@math.qau.edu.pk

Abstract: Carbon-fiber composites are considered to be one of the suitable materials for the fabrication of prosthetic feet. However, commercially available composites-based prosthetic foot designs present several problems for lower limb amputees, such as low tensile strength, reduced impact resistance, high cost, and weight structure. Modulating the mechanical properties of carbon-fiber composites using a simplified method can help reduce these issues. Therefore, our present research aims to identify the impact of increasing the concentration of carbon fiber in the fabrication of carbon-fiber composites by using the hand layup method without the vacuum bagging technique. To improve the mechanical strength of carbon-fiber laminates, an increasing number of carbon-fiber layers are used in sample preparation. This study aims to determine the tensile strength of the laminates with a different number of carbon-fiber laminations. For the preparation of the sample specimen, black 100% 3 K 200 gsm carbon fiber with a cloth thickness of 0.2 mm and tensile strength of 4380 Mpa was laminated with two parts of epoxy resin Araldite® LY556 and Aradur hardener at a ratio of 100:30 to make the test specimen. The results indicated an overall improvement in the tensile strength of carbon-fiber laminates owing to the increase in the number of carbon-fiber layers in successive samples. The maximum achieved tensile strength through the present experimental protocol is 576.079 N/mm^2, depicted by a prepared specimen of 10 layers of carbon fiber. Secondly, an increase in the deformation rate has also been observed by increasing the loading rate from 2 mm/min to 5 mm/min during the tensile testing of fabricated samples. These sample carbon-fiber composites can be used in the fabrication of prosthetic feet by controlling the experimental conditions. The fabricated prosthetic foot will assist in rehabilitating lower-limb amputees.

Keywords: carbon-fiber composites; tensile strength; deformation rate; prosthetic foot

Citation: Sehar, B.; Waris, A.; Gilani, S.O.; Ansari, U.; Mushtaq, S.; Khan, N.B.; Jameel, M.; Khan, M.I.; Bafakeeh, O.T.; Tag-ElDin, E.S.M. The Impact of Laminations on the Mechanical Strength of Carbon-Fiber Composites for Prosthetic Foot Fabrication. *Crystals* **2022**, *12*, 1429. https://doi.org/10.3390/cryst12101429

Academic Editor: Pavel Lukáč

Received: 10 September 2022
Accepted: 30 September 2022
Published: 10 October 2022

1. Introduction

Lower-limb amputations severely affect quality of life by restraining body functions such as movement. Lower-limb prosthetics rehabilitate not only mobility but also the well-being of amputees. One of the primary factors in rehabilitating lower limb amputees is the prosthetic foot. For this purpose, several prosthetic foot designs have been introduced since the SACH (Solid-Ankle Cushion-Heel) foot was developed in 1957 [1]. SACH has been prescribed to disabled patients because it could lower the impact loading at heel strike.

Nevertheless, this commonly used prosthetic foot can store and release a small amount of elastic energy [2]. Other earlier prosthetic foot designs consisting of wood, metal, and vulcanized rubber presented various issues for amputees, such as lack of durability and discomfort. Previously, problems with prosthetic foot designs, including the Shape and Roll foot, Niagara Foot, and Jaipur foot, were investigated by several studies [3–5], stating similar issues of reduced durability and increased weight. Owing to problems with formerly used prosthetic foot materials, carbon-fiber composites were considered a suitable alternative, as stated by previous research [6]. Nolan [7] supported using composite materials for designing prosthetic feet because of their energy storage capacity, which provides a rehabilitative advantage to amputees requiring a high activity level.

Similarly, Campbell [8] also supported using carbon fiber and glass fiber in prosthetic applications for their low density, lightweight quality, and high strength. Recently, another study in Vietnam also provided evidence that carbon-fiber laminated prosthetic feet store elastic energy, helping the body move forward and reducing the impact force on residual limbs [9]. This property increases the flexibility of the prosthetic foot, thus mimicking the working of a natural foot.

Carbon fiber has been a primary choice for industrial uses for over a century. Their excellent strength-to-weight and stiffness-to-weight ratios make them suitable for the automotive, aerospace, and prosthetic industries. According to Sau-Fun et al. [10], previously, wood and metal were utilized as the preferred materials for constructing and fabricating prosthetic devices. However, Maruo et al. [11] suggested that such prosthetic materials presented issues such as reduced resistance to moisture and corrosion, and dampness. Similarly, another study stated that prosthetic applications require durable and stable materials such as synthetic fibers to provide comfort and control to amputees. Scholz et al. [12] related the supremacy of these composite materials over other materials to their biocompatibility and strength-to-weight characteristics.

Furthermore, Liu et al. [13] also claimed that the enhanced characteristics of fiber composite materials make them suitable for the prosthetic industry and industries such as sports, aeronautics, and aerospace. In recent studies, various other forms of composites, such as polypropylene (PP)-based nanocomposites, have also been tempered with nanoclay, basalt fibers, and graphene to be used in the aerospace industry [14,15]. Another form of composites known as fiber laminated metals (FML) has also been studied to investigate their impact on the mechanical strength of aircraft components. The results depicted an overall increase in tensile strength using fiber laminated metals [16]. Modulating fiber composites with different techniques can enhance their mechanical properties, making them suitable for various manufacturing industries [17].

Prosthetic utilization has been significantly dependent on carbon fibers due to their ability to store energy and their adaptability, according to Dziaduszewska et al. [18]. The properties of carbon fibers can be manipulated by fabricating carbon-reinforced composites using epoxy resins and woven carbon fibers. Oleiwi et al. [19] explained that the lamination process could incorporate specific tensile properties and rigidity by controlling the specific angles and altering the matrix. Using various lamination techniques, blended polymer frameworks can be constructed to be utilized for the manufacturing of prosthetic limbs, as depicted by an experimental protocol designed by Abbas et al. [20]. Nowadays, the rapid prototyping system uses metal, plastic, or other material laminations to fabricate orthotic and prosthetic devices [21].

Several theoretical and experimental studies have been carried out to improve the mechanical properties of carbon-fiber composites. Jweeg et al. [22] have characterized two different fiber-type composite materials resulting in better performance outcomes for unidirectional fibers. Similarly, Abbas et al. [20] supported the improved mechanical properties of carbon-fiber composites by increasing the number of layers in carbon-fiber laminates during the manufacturing process. A recent study conducted by Khare et al. [23] also depicted that flexural strength increases by 17% when using 15% carbon-fiber reinforcements. Likewise, Rahmani et al. [24] supported the fact that the mechanical strength in composites can be improved by manipulating the interlaminar properties and strengthening the resin matrix. Hadi et al. [25] demonstrated a greater impact of the reinforcement material on improved tensile strength rather than the direction of polymer laminates. Muhammed et al. [26] suggested a relationship between the tensile strength and weight percentage of the reinforcement material that enhances certain properties, including elasticity, yield strength, and ultimate strength, by changing the weight percentage of the fiber composites.

Over previous decades, several cost-effective techniques were presented for the fabrication of carbon-fiber composites to improve their mechanical properties. Such techniques include hand layup, compression molding, vacuum bagging, and vacuum-assisted resin transfer molding to manufacture epoxy-based composites, as described by Pulikkalparambil et al. [27]. The improvement of prepared composites' properties depends on the suitability and efficacy of the selected technique. For instance, Muralidhara et al. [28] previously developed carbon-fiber laminates using the hand layup method followed by vacuum bagging, producing a significant improvement in the mechanical performance of the carbon-fiber composites. However, the tensile and flexural strength of the material was significantly enhanced by vacuum bagging compared to the hand layup method. Chen et al. [29] provided a different perspective regarding the wet hand layup method's limitations compared to fiber 3D printing (F3DP). The study concluded that the hand layup method fabricated prostheses with great flexure rather than tensile strength.

In this work, a simplified hand layup method has been utilized to fabricate carbon-fiber laminates to test the suitability of carbon fibers for prosthetic purposes. Carbon-fiber layers were subsequently increased in the prepared carbon-fiber laminates to test the impact of increasing the fiber content over the mechanical strength. The previous study conducted by Al-Khazraji et al. [30] depicts the suitability of such laminates for prosthetic sockets. The present study will focus on the suitability of such materials for designing a prosthetic foot prototype.

2. Materials and Methods

2.1. Materials

To test the mechanical strength of carbon-fiber laminates, the following materials have been used for the preparation of samples:

1. Black 100% 3 K 200 gsm carbon-fiber cloth with tensile strength 4380 Mpa (thickness = 0.2 mm);
2. Two parts epoxy resin Araldite® LY556;
3. Hardener Aradur® 22962;
4. Mold releasing wax;
5. Plywood for mold preparation.

2.2. Experiment Protocol

Specimen Preparation

A 100% 3 K 200 gsm carbon-fiber cloth with a thickness of 0.2 mm was cut into dimensions of 24 × 2.5 mm to prepare several layers of carbon-fiber lamination. To prepare the required samples of laminates for tensile testing, the carbon-fiber cloth with 0.2 mm thickness was stacked to up to 10 layers. For the preparation of each sample, 2, 6, and 10 carbon fabric layers were laminated using epoxy resin in a ratio of 100:30 by volume. The thickness of each sample varied owing to the difference in the number of carbon-fiber laminations. In a drying oven, these laminated samples were cured at a high temperature of between 80 °C and 120 °C.

For the preparation of carbon-fiber laminates, the fibers were kept at the orientation of 0° and 90°. Further, for constructing the laminates, two parts of an epoxy resin with the properties listed in Table 1 were applied using a paintbrush to acquire an equal distribution of resin between the subsequent layers of carbon fiber.

The hand layup method was used to fabricate the required carbon-fiber samples and the wooden molds to assist in curing the samples. The hand layup process is depicted in Figure 1. The hand layup method is one of the most significant processes in the composite manufacturing industry. This process helps build high-performance composite materials with enhanced mechanical as well as structural properties [31]. Carbon-fiber composites are fabricated layer by layer and incorporated with epoxy resin. The hand layup process consists of carbon-fiber composites, including layers of well-aligned fiber reinforcements with the matrix material. The prepared carbon-fiber laminates can be used for various applications, including sports goods, automotive, and prostheses. For the fabrication of a prosthetic foot prototype, the first step is the design construction using SolidWorks 2017 software. Three parts of the prosthetic foot (sole, keel, and heel) are designed using the approximate measurements of an average foot size (24.5 cm).

Table 1. Properties of epoxy resin and hardener.

Properties	Epoxy Resin Araldite® LY556	Hardener Aradur® 22962
Viscosity at 25 °C (ISO 12058–1)	10,000–12,000 mPa s	5–20 mPa s
Density at 25 °C (ISO 1675)	1.15–1.20 g/cm^3	0.89–0.90 g/cm^3
Flash Point (ISO 2719)	>200 °C	≥110 °C
Storage Temperature	2–40 °C	2–40 °C
Epoxy content (ISO 3000)	5.30–5.45 eq/kg	——
Epoxy equivalent (ISO 3000)	183–189 g/eq	——

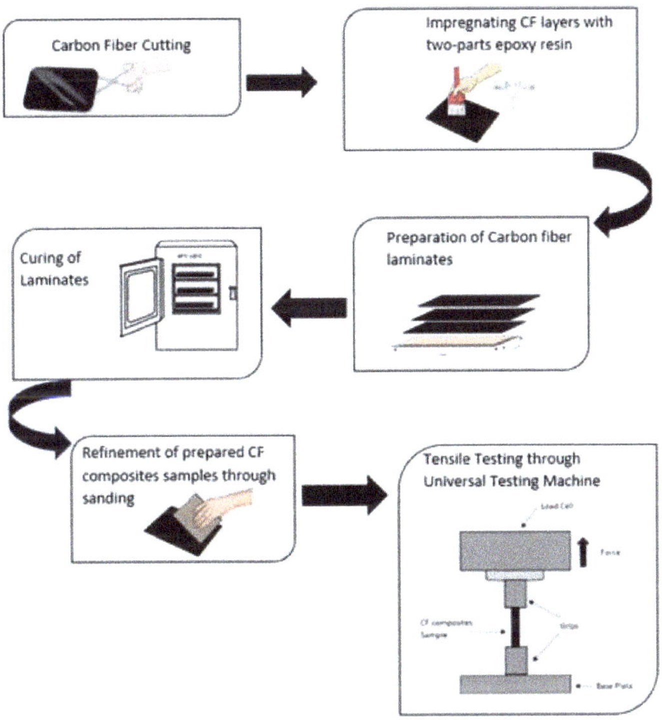

Figure 1. Hand Layup Process for producing carbon-fiber composite samples.

Laminates are cured in the drying oven at temperatures of 80 °C to 120 °C. The prepared samples are then subjected to shear cutting and sanding to refine the laminated samples. The process is repeated for several samples to fabricate the required number of samples used in the universal testing machine. Table 2 depicts the experimental protocol for sample fabrication.

Table 2. Experimental protocol.

Araldite® LY556/Aradur® 22962	Number of CF Laminations	Load Rate
100:30	2 layers 4 layers	2 mm/min, 2 mm/min
	2 layers	5 mm/min
	6 layers	5 mm/min
	10 layers	5 mm/min

2.3. Tensile Testing

Tensile testing is one of the crucial steps in determining the mechanical strength of a particular material, which is carbon-fiber laminates in the present case. A universal testing machine is used to apply a uniaxial load on the sample until the point of failure, as shown in Figure 2. The acquired result from the tensile testing can be utilized to select suitable materials for quality control or prosthetic applications. The mechanical testing also predicts the properties of the material being tested. These properties include maximum elongation or reduction in the cross-sectional area and ultimate strength. For this purpose, the specimen is loaded in a controlled manner in the grip section of the universal testing machine (UTM). The materials are prepared and tested according to the ASTM D3039 standard, which provides the guidelines for testing the laminated type of composite material [32]. The maximum capacity of the universal testing machine is 20 KN, which worked on the electronic control servo mechanism. The gauge length for each sample was kept at approximately 80 mm. Furthermore, the material was subjected to tensile testing at 5 mm/min.

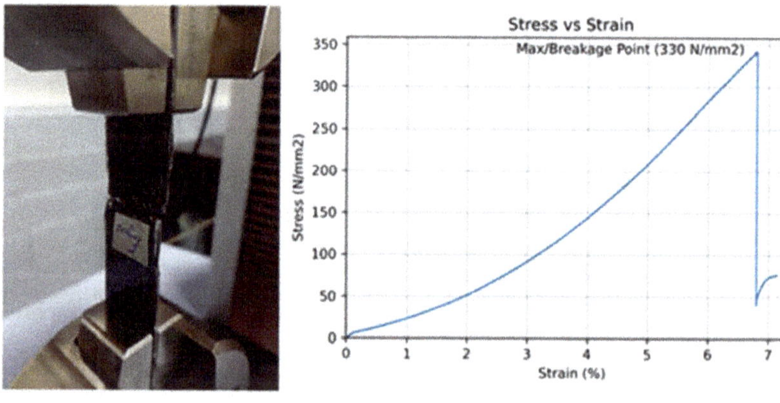

(a) (b)

Figure 2. (**a**) Tensile Test for carbon-fiber composites in UTM (**b**) Mechanical behavior of carbon-fiber composites in UTM.

2.4. Design Concept of Prosthetic Foot

For the fabrication of a prosthetic foot prototype, the first step is the design construction using SolidWorks 2017 CAD software (version 25, Dassault Systèmes, Waltham, MA, USA). Three parts of the prosthetic foot (sole, keel, and heel) are designed using the approximate measurements of an average foot size (24.5 cm). The sole, keel, and heel dimensions are 24.5 cm × 9 cm, 19.2 cm × 7 cm, and 8.3 cm × 5 cm, respectively. The thickness of each part was kept equal to 0.7 cm, as the prosthetic footplates were manufactured using the same number of carbon-fiber layers. Figure 3 illustrates the CAD design and fabricated prosthetic foot prototype.

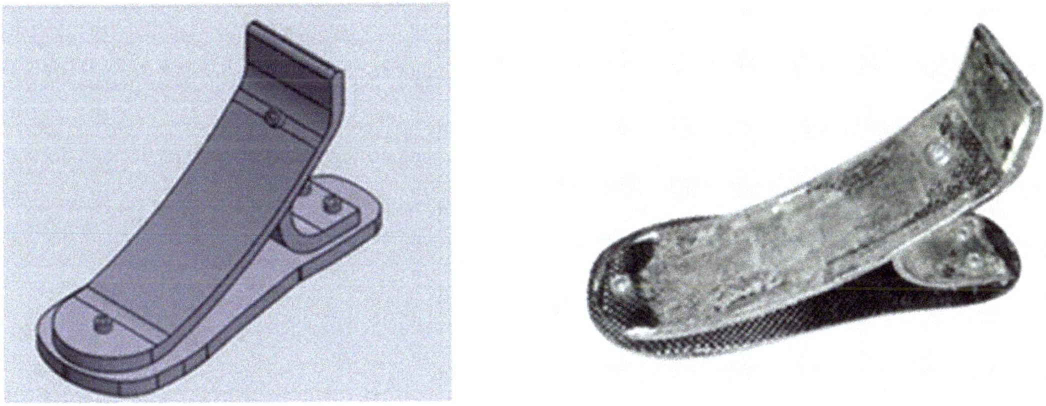

(a) (b)

Figure 3. (**a**) CAD design for prosthetic foot prototype (**b**) Fabricated prosthetic foot prototype.

3. Results

The mechanical strength of three specimens of carbon-fiber laminates were tested using the universal testing machine. The first sample was prepared using two carbon-fiber layer reinforcements (0°/90°) and epoxy resin in a ratio of 100:30 by volume and tested to determine the maximum stress capacity. This two-layered sample exhibited a mechanical strength of 251 N/mm^2 before the failure of the carbon-fibers in the sample. Figure 4a depicts the ultimate tensile strength along with the failure of the tested sample.

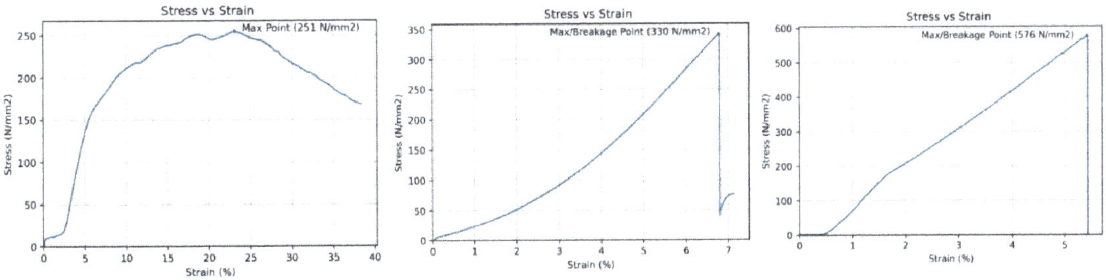

(a) (b) (c)

Figure 4. (**a**) Stress–strain Curve for Sample 1 (**b**) Stress–strain Curve for Sample 2 (**c**) Stress–strain Curve for Sample 3.

Similarly, the second laminated sample was prepared using the six layers of carbon fibers with the same ratio of two parts of epoxy resin. The increase in the number of CF layers increased the stress-enduring capability of the laminate. Table 3 shows the parameters obtained when tensile testing the carbon-fiber laminates along with the stress–strain curve generated by the universal testing machine. The stress–strain graph in Figure 4b shows a higher ultimate tensile strength for the sample in comparison to the previous one in Figure 4a. Secondly, the material breaks after reaching the yield point, which suggests the maximum load-bearing capability of the fibers. For this sample, 10 layers of carbon fiber were added along with the 100:30 of two parts epoxy resin. The mechanical properties of the sample are evident from the data and stress–strain curve in Figure 4c.

Table 3. UTM Generated Data for Carbon-fiber Composites.

Sample	Time (sec)	Stress (N/mm^2)	Strain (%)	Force (N)	Displacement (mm)	Stroke (mm)
Sample 1	0	−0.06769	0	−2.26498	0	0
	0.1	−0.13361	0.044444	−0.54995	0.005333	0.005333
	27.23	251.2207	18.88733	1034.025	2.266479	2.266479
	45.36	208.7001	31.47639	859.0095	3.777167	3.777167
	55.26	167.9828	38.35209	691.4171	4.60225	4.60225
Sample 2	0	−2.85135	0	−98.3715	0	0
	0.1	−2.77994	0.002438	−95.9078	0.002146	0.002146
	179.64	330.9781	6.805351	11,418.74	5.988708	5.988708
	188.58	75.94242	7.141951	2620.013	6.284916	6.284916
	188.4	75.85056	7.135014	2616.844	6.278812	6.278812
Sample 3	0	−0.06769	0	−2.26498	0	0
	0.1	0.266715	0.006849	8.924802	0.005479	0.005479
	52.1	576.079	5.426823	19276.75	4.341458	4.341458
	52.12	540.9233	5.429844	18100.37	4.34z3875	4.343875
	52.18	−0.29973	5.432031	−10.0295	4.345625	4.345625

Several samples showed abnormal behavior under the application of force by the universal testing apparatus. Figure 5a,b illustrate the stress–strain curve of the failure of several laminated samples. Most of these samples were tested at a lower load rate which resulted in different tensile behavior. Figure 5a,b depicts the stress–strain graph for the samples tested under a load rate of 2 mm/min in the Universal Testing Machine.

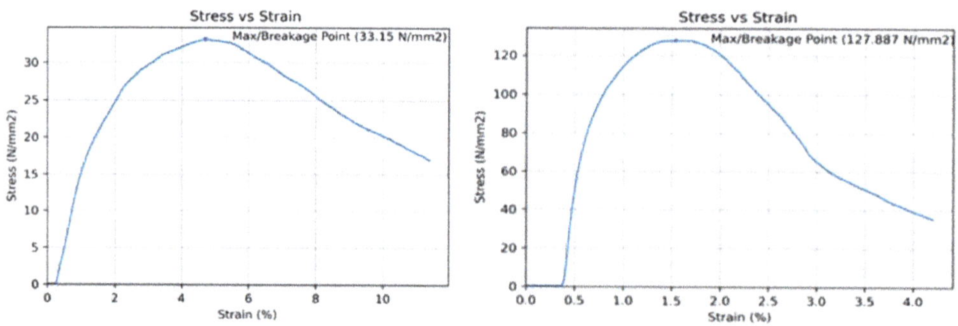

a **b**

Figure 5. (**a**) Stress–strain curve for two-layered sample (Failed Sample 1) (**b**) Stress–strain curve for four-layered sample (Failed Sample 2).

Prosthetic Foot Prototype Results

The prototype for the prosthetic foot was tested under a static load of 68,948 N/m^2 to check the displacement of footplates. Figure 6a shows the displacement under a load of zero. However, Figure 6b shows the displacement of footplates when a static load (68,498 N/m^2) was applied for a few minutes. The displacement recorded under the load is 10 mm in the heel section of the foot. The static loading on the fabricated prosthetic foot was performed to assess the stability. Various other methods such as cyclic loading, MTS testing, and ANSYS Static Structural Analysis can be used for further investigation of a prosthetic foot in terms of its range of motion [33]. However, due to the lack of the required apparatus for cycling loading, mechanical testing was performed using a static load.

(a) (b)

Figure 6. (**a**) Initial 0 mm displacement in the heel section; (**b**) 10 mm displacement under static load.

4. Discussion

In the present study, tensile testing of carbon-fiber laminates indicates that increasing carbon-fiber content for composite fabrication can enhance mechanical strength. The mechanical properties of different samples are depicted in Table 4. Furthermore, the stress–strain curves generated by the universal testing machine can help to determine the mechanical properties of the used material. The stress–strain curve shows a linear behavior at certain points and non-linear curvatures at the start of the mechanical test. The linear sections of these graphs indicate that increasing the load on the laminates increases the strain rate along with a deformation of the fibers present in the sample, resulting in breakage of the tested carbon-fiber laminates. It was suggested by Wang et al. [34] that the failure of composites occurs when carbon fibers in the samples are fractured. The failure points in the reported study were observed at 470 MPa and 800 MPa, respectively. These failure stress points show the functional failure point of the composite sample.

Table 4. Mechanical Strength of Tested Carbon-fiber Laminates.

CF Layers	Load Rate	Epoxy: Resin	Mechanical Strength (N/mm^2)
2	2 mm/min	100:30	33.15
4	2 mm/min	100:30	127.88
2	5 mm/min	100:30	254.51
6	5 mm/min	100:30	341.54
10	5 mm/min	100:30	576.07

However, the non-linear regions in the stress–strain curve indicate the viscoelastic nature of the matrix, along with the brittle failure of the composite sample. Secondly, the slope in the stress–strain graphs show the initial stiffness of the carbon-fiber composites. The lowest tensile strength is shown by the two-layered laminates tested at a loading rate of 2 mm/min. To improve the tensile strength, other two-layered laminates were tested at an increased loading rate of 5 mm/min. Khan et al. [35] have reported a direct dependency of fracture strain and stress on the load rate in the case of carbon-fiber composites.

The increase in the number of layers in successive samples indicated a direct impact on the tensile strength of the carbon-fiber laminates, as evident from Figure 7. The tensile strength is mainly affected by the type of fiber used in the composition of the laminate. Furthermore, the applied matrix also varies the properties of the composite by improving the adhesion bond between carbon-fiber layers [28]. The mechanical properties of the composite materials also depend on the fabrication technique.

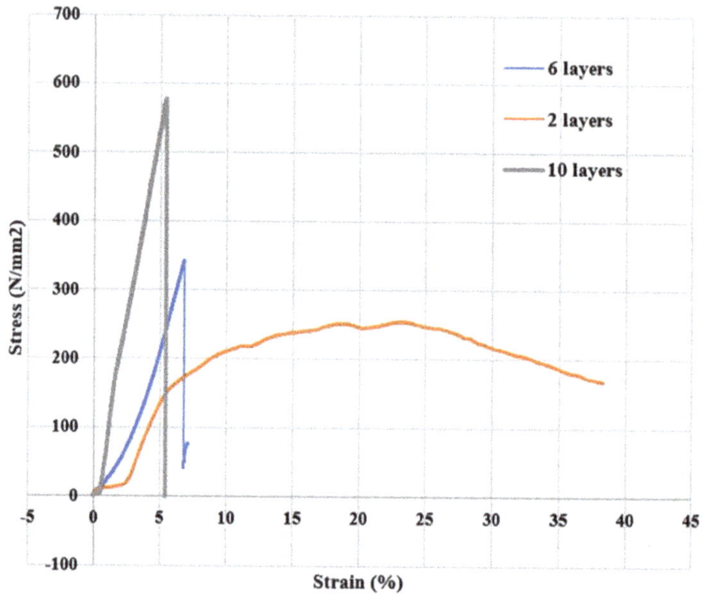

Figure 7. Stress–strain Superimposed scattered plot for tested carbon-fiber samples.

As evident from the stress–strain curves, the yield point (maximum point for linear behavior) is followed by the tensile deformation of the sample resulting from the failure of fibers under stress. The cross plies (0°/90°) are the major regions initiating the cracks in the composite samples. The breakage occurs in those areas in fiber structure which have a lower tensile strength than the ideal tensile strength of the unidirectional composites. For instance, one of these breakage points is depicted in the stress–strain graph for sample 2, which is at 351.54 N/mm^2. Moreover, failure of the reinforced fiber composites can also occur due to the formation of hollow spaces and the uneven distribution of the carbon-fibers in the epoxy matrix. Another issue in the failure of the mechanical testing is the slippage of the fibers in the grip section of the UTM. Concerning the reduced tensile strength, Khan et al. [35] have reported a direct dependency of tensile strength over the loading rate. Our study supported this dependency by depicting a reduced tensile strength and failure behavior at a loading rate of 2 mm/min, as illustrated in Figure 5a,b.

The deflection of the heel section under a high static load (68,948 N/m^2) shows that the prosthetic foot can withstand higher stress conditions. The minimum deflection of 10 mm also shows the lower probability of delamination of the fabricated carbon-fiber plates under static load. The tested carbon-fiber laminates with high mechanical strength can be used for multiple prosthetic and orthotic purposes.

Our research depicted different behaviors in the mechanical strength of the woven carbon-fiber composites than previously reported. Our study also simplified the overall manual process of the hand layup method by eliminating the vacuum bagging process, which was a commonly used procedure in previous studies, depicted in Table 5. Using the hand layup method can increase the probability of forming an uneven matrix and fiber distribution, along with hollow sections which can lead to the failure of carbon-fiber composites. The difference in mechanical behavior and non-linearity can be due to factors such as experimental settings and manufacturing techniques. Such factors must be modulated to improve the tensile behavior of carbon-fiber composites in future studies. Furthermore, carbon-fiber composites are considered an expensive material for prosthetic applications, which can be a limitation in manufacturing carbon-fiber composites on an industrial scale.

Table 5. Comparison of Findings of Previous Studies.

Previous Studies	Materials	Fabrication Method	Tensile Strength	Conclusion
(Muralidhara et al. 2020)	Carbon fiber: T800CF/Ep, T700CF/Ep, and T300CF/Ep Epoxy Resin: Araldite LY1564 Hardener: Aradur 22962	Hand layup method with vacuum bagging process	Approximately 680 MPa, 630 MPa, and 330 Mpa.	(2–6) % increase in the mechanical strength by vacuum bagging in comparison to the hand layup method.
(Chen et al. 2021)	Unidirectional carbon-fiber sheets Two parts epoxy resin	Hand layup method	The mean tensile strength of 13 CF samples showed an average tensile strength equal to 164.57.	The hand layup method provided higher stiffness and mechanical strength in flexure.
(Pham et al. 2020)	Dry carbon-fiber fabric Polyester resin	Hand layup method with vacuum bagging	Specimen tensile strength was found to be 243 Mpa.	The manufactured prosthetic foot prototype will enable forward propulsion lowering the impact force upon residual organs.
[36] (Karthik et al. 2021)	Glass, carbon, and Kevlar fibers	Hand layup method with compression molding	A mixture of carbon and Kevlar fibers indicated the highest tensile strength of 385.09 Mpa.	Carbon-Kevlar-Carbon composites showed fewer surface defects under stress.

5. Conclusions

In this research, the hand layup process was used with no vacuum bagging apparatus to investigate the impact of increasing carbon-fiber content on the mechanical strength of carbon-fiber composites. Through the optimization of this process, carbon-fiber composite plates were fabricated, which were used to construct a prosthetic foot prototype. Mechanical testing of various layered carbon-fiber composites under different loading rates (2 mm/min and 5 mm/min) depicted different tensile strengths. The minimum number of carbon-fiber layers (n = 2) demonstrated the lowest tensile strength of 254.51 N/mm^2. Subsequently, a slight improvement in the mechanical strength was observed in the six-layered composite sample, demonstrating a tensile strength of 341.54 N/mm^2. However, the ten-layered sample had the highest tensile strength of equal to 576.07 N/mm^2. The Universal Testing Machine results suggested that increasing the number of CF laminations can improve mechanical strength. The acquired results were different from the previously reported data in terms of having lower tensile strength. The difference in the tensile behavior might be the result of different experimental procedures. The hand-layup method is effective in improving flexural strength and stiffness of the material rather than tensile strength as evidenced by previous studies reported in this study. Secondly, the breakage of the specimen normally occurs at the cross plies after reaching the yield point, indicating the maximum stress that the specimen can withstand. The fracture behavior depicted the brittle behavior of the tested material. By controlling the environmental conditions and experimental parameters in future studies, the mechanical strength of the prepared composite samples can be further enhanced.

Author Contributions: Conceptualization: B.S., A.W., S.O.G., U.A., S.M., N.B.K., M.J., M.I.K., O.T.B. and E.S.M.T.-E.; data curation: B.S., A.W., S.O.G. amd U.A.; formal analysis: B.S., A.W. and U.A.; funding acquisition: research received no external funding; investigation: B.S., A.W. and U.A.; methodology: B.S., A.W. and U.A.; project administration: B.S., A.W. and U.A.; resources: A.W., S.O.G. and U.A.; software: B.S., A.W. and U.A.; supervision: A.W., S.O.G. and U.A.; validation: A.W., S.O.G. and U.A.; visualization: B.S. and A.W.; writing—original draft: B.S.; review and editing: B.S., A.W., S.O.G., U.A., S.M., N.B.K., M.J., M.I.K., O.T.B. and E.S.M.T.-E. All authors have read and agreed to the published version of the manuscript.

Funding: This research received no external funding.

Institutional Review Board Statement: Not Applicable.

Informed Consent Statement: Not Applicable.

Data Availability Statement: All data generated or analyzed during this study are included in this published article.

Acknowledgments: The authors extend their appreciation to the deanship of scientific research at King Khalid University for funding this work through large group project under grant number (RGP. 2/93/43).

Conflicts of Interest: The authors declare no conflict of interest

References

1. Bhargava, R. The Jaipur Foot and the 'Jaipur Prosthesis'. *Indian J. Orthop.* **2019**, *53*, 5. [CrossRef]
2. South, B.J.; Fey, N.P.; Bosker, G.; Neptune, R.R. Manufacture of Energy Storage and Return Prosthetic Feet Using Selective Laser Sintering. *J. Biomech. Eng.* **2009**, *132*, 015001. [CrossRef]
3. Meier, M.R. Design and development of appropriate technology for low-income countries: Preliminary field investigation of the Shape & Roll prosthetic foot in El Salvador, Central America. *Capabilities* **2002**, *11*, 1–2.
4. Hahl, J.; Taya, M. Experimental and numerical predictions of the ultimate strength of a low-cost composite transtibial prosthesis. *J. Rehabil. Res. Dev.* **2000**, *37*, 405–413. [PubMed]
5. Arya, A.P.; Klenerman, L. The Jaipur foot. *J. Bone Jt. Surgery. Br. Vol.* **2008**, *90*, 1414–1421. [CrossRef]
6. Jagannatha, T.D.; Harish, G. Mechanical properties of carbon/glass fiber reinforced epoxy hybrid polymer composites. *Int. J. Mech. Eng. Robot. Res.* **2015**, *4*, 131–137.
7. Nolan, L. Carbon fibre prostheses and running in amputees: A review. *Foot Ankle Surg.* **2008**, *14*, 125–129. [CrossRef] [PubMed]
8. Campbell, F.C. *Structural Composite Materials*; Google Books; ASM International: Materials Park, OH, USA, 2010.
9. Pham, H.T.; Phan, T.V.; Mai, V.T. Optimization Design of a Carbon Fibre Prosthetic Foot for Amputee. *Acta Sci. Orthop.* **2020**, *3*, 16–21. [CrossRef]
10. Sau-Fun, N.; Chi-Leung, H.; Lai-Fan, W. Development of medical garments and apparel for the elderly and the disabled. *Textile Progress* **2011**, *43*, 235–285. [CrossRef]
11. Maruo, Y.; Nishigawa, G.; Irie, M.; Yoshihara, K.; Minagi, S. Flexural properties of polyethylene, glass and carbon fiber-reinforced resin composites for prosthetic frameworks. *Acta Odontol. Scand.* **2015**, *73*, 581–587. [CrossRef]
12. Scholz, M.-S.; Blanchfield, J.; Bloom, L.; Coburn, B.; Elkington, M.; Fuller, J.; Gilbert, M.; Muflahi, S.; Pernice, M.; Rae, S.; et al. The use of composite materials in modern orthopaedic medicine and prosthetic devices: A review. *Compos. Sci. Technol.* **2011**, *71*, 1791–1803. [CrossRef]
13. Liu, L.; Jia, C.; He, J.; Zhao, F.; Fan, D.; Xing, L.; Wang, M.; Wang, F.; Jiang, Z.; Huang, Y. Interfacial characterization, control and modification of carbon fiber reinforced polymer composites. *Compos. Sci. Technol.* **2015**, *121*, 56–72. [CrossRef]
14. Wang, J.; Song, F.; Ding, Y.; Shao, M. The incorporation of graphene to enhance mechanical properties of polypropylene self-reinforced polymer composites. *Mater. Des.* **2020**, *195*, p. 109073. [CrossRef]
15. Thongchom, C.; Refahati, N.; Saffari, P.R.; Saffari, P.R.; Niyaraki, M.N.; Sirimontree, S.; Keawsawasvong, S. An Experimental Study on the Effect of Nanomaterials and Fibers on the Mechanical Properties of Polymer Composites. *Buildings* **2021**, *12*, 7. [CrossRef]
16. Khalid, M.Y.; Arif, Z.U.; Ahmed, W.; Arshad, H. Evaluation of tensile properties of fiber metal laminates under different strain rates. *Proc. Inst. Mech. Eng. Part E J. Process. Mech. Eng.* **2022**, *236*, 556–564. [CrossRef]
17. Bheel, N.; Tafsirojjaman, T.; Liu, Y.; Awoyera, P.; Kumar, A.; Keerio, M.A. Experimental Study on Engineering Properties of Cement Concrete Reinforced with Nylon and Jute Fibers. *Buildings* **2021**, *11*, 454. [CrossRef]
18. Dziaduszewska, M.; Wekwejt, M. Composites in energy storing prosthetic feet. *Eur. J. Med. Technol.* **2018**, *3*, 16–22.
19. Oleiwi, J.K.; Hadi, A.N. Experimental and numerical investigation of lower limb prosthetic foot made from composite polymer blends. *Int. J. Mech. Prod. Eng. Res. Dev.* **2018**, *8*, 1319–1330.
20. Abbas, S.M.; Resan, K.K.; Muhammad, A.K.; Al-Waily, M. Mechanical and Fatigue Behaviors of Prosthetic for Partial Foot Amputation with Various Composite Materials Types Effect. *Int. J. Mech. Eng. Technol. (IJMET)* **2018**, *9*, 383–394.
21. Barrios-Muriel, J.; Romero-Sánchez, F.; Alonso-Sánchez, F.J.; Salgado, D.R. Advances in Orthotic and Prosthetic Manufacturing: A Technology Review. *Materials* **2020**, *13*, 295. [CrossRef]
22. Jweeg, M.J.; Hammood, A.S.; Al-Waily, M. Experimental and theoretical studies of mechanical properties for reinforcement fiber types of composite materials. *Int. J. Mech. Mechatron. Eng.* **2012**, *12*, 62.
23. Khare, J.M.; Dahiya, S.; Gangil, B.; Ranakoti, L. Influence of different resins on Physico-Mechanical properties of hybrid fiber reinforced polymer composites used in human prosthetics. *Mater. Today: Proc.* **2020**, *38*, 345–349. [CrossRef]
24. Rahmani, H.; Najafi, S.H.M.; Ashori, A. Mechanical performance of epoxy/carbon fiber laminated composites. *J. Reinf. Plast. Compos.* **2014**, *33*, 733–740. [CrossRef]
25. Hadi, A.N.; Oleiwi, J.K. Improve Flexural Strength of PMMA/SR Polymer Blend by Reinforcement with Carbon Fibers as Prosthetic Foot Polymer Material. *Int. J. Appl. Or Innov. Eng. Manag.* **2015**, *4*, 172–176.

26. Muhammed, A.M. Experimental investigation of tensile and fatigue stresses for orthotic/prosthetic composite materials with varying fiber (perlon, e-glass and carbon). *ARPN J. Eng. Appl. Sci.* **2016**, *11*, 12820–12827.
27. Pulikkalparambil, H.; Rangappa, S.M.; Siengchin, S.; Parameswaranpillai, J. Introduction to Epoxy Composites. In *Epoxy Composites*, 1st ed.; Parameswaranpillai, J., Pulikkalparambil, H., Rangappa, S.M., Siengchin, S., Eds.; Wiley: Weinheim Germany, 2021; pp. 1–21. [CrossRef]
28. Muralidhara, B.; Babu, S.K.; Suresha, B. Utilizing vacuum bagging process to prepare carbon fiber/epoxy composites with improved mechanical properties. *Mater. Today: Proc.* **2020**, *27*, 2022–2028. [CrossRef]
29. Chen, A.Y.; Baehr, S.; Turner, A.; Zhang, Z.; Gu, G.X. Carbon-fiber reinforced polymer composites: A comparison of manufacturing methods on mechanical properties. *Int. J. Light. Mater. Manuf.* **2021**, *4*, 468–479. [CrossRef]
30. Al-Khazraji, K.; Kadhim, J.; Ahmed, P.S. Tensile and fatigue characteristics of lower-limb prosthetic socket made from composite materials. In Proceedings of the 2012 International Conference on Industrial Engineering and Operations Management, Istanbul, Turkey, 3–6 July 2012; pp. 847–852.
31. Elkington, M.; Bloom, D.; Ward, C.; Chatzimichali, A.; Potter, K. Hand layup: Understanding the manual process. *Adv. Manuf. Polym. Compos. Sci.* **2015**, *1*, 138–151. [CrossRef]
32. de Paiva, J.M.F.; Santos, A.D.N.D.; Rezende, M.C. Mechanical and morphological characterizations of carbon fiber fabric reinforced epoxy composites used in aeronautical field. *Mater. Res.* **2009**, *12*, 367–374. [CrossRef]
33. Haislip, T.S. Comparative Analysis Of Two Prosthetic Foot Designs Using Cyclical Load Testing And Cad Simulation. Ph.D. Thesis, Mercer University, Macon, GA, USA, 2019.
34. Wang, J.; Chen, L.; Shen, W.; Zhu, L. Research on Tensile Properties of Carbon Fiber Composite Laminates. *Polymers* **2022**, *14*, 2318. [CrossRef]
35. Khan, H.A.; Nigar, M.; Chaudhry, I.A. Tensile Behavior of Unidirectional Carbon Reinforced Composites for Aerospace Structures under Varying Strain Rates. *Appl. Mech. Mater.* **2015**, *798*, 357–361. [CrossRef]
36. Karthik, K.; Rajamani, D.; Manimaran, A.; Udayaprakash, J. Evaluation of tensile properties on Glass/Carbon/Kevlar fiber reinforced hybrid composites. *Mater. Today: Proc.* **2020**, *39*, 1655–1660. [CrossRef]

MDPI
St. Alban-Anlage 66
4052 Basel
Switzerland
www.mdpi.com

Crystals Editorial Office
E-mail: crystals@mdpi.com
www.mdpi.com/journal/crystals

www.ingramcontent.com/pod-product-compliance
Lightning Source LLC
Chambersburg PA
CBHW062103220926
51850CB00010B/2871